Solar Power

Design Manual

Second Edition

Richard A Stubbs

Solar Power Design Manual

Contents

www.solar-power-answers.co.uk

1 Introduction

1.1 Scope

This book is intended to give the reader sufficient knowledge to design and install a stand-alone solar power system anywhere in the world. It covers the principles of photovoltaic power generation and energy conversion and goes on to outline the necessary design and installation procedures.

Grid-connected solar systems are not covered in great detail, as they can not be considered to be 'stand-alone', however basic information on their function is included for information. Certain elements of system design are applicable to grid-connected systems, and the reader is made aware of where this is the case.

The resources required are included where necessary and there are illustrations as appropriate. It is recommended that you read the entire book before attempting any of the procedures within.

1.2 Experience

The reader is assumed to have a certain amount of knowledge and previous experience including basic electrical and mechanical knowledge. Experience of common tools will be an advantage. Some calculations are required although every attempt has been made to make the process of system design as simple as possible.

1.3 Disclaimer

Every care has been taken to ensure that the information contained in this book is correct. However, it is based on personal experience and may not be applicable to every situation. No responsibility is accepted for any loss suffered, either directly or indirectly, as a result of the information contained in this eBook.

2 Basic Principles

2.1 Volts, Amps and Watts

Throughout this book there are references to Voltage, Current, Power and Resistance. It is important to understand what each of these means and how they relate to each other. The units for each are:

- Voltage: The potential difference between two points. Is measured in Volts (V) and has the symbol 'V'.

- Current: The flow of electrons in a circuit. Is measured in Amps (A) and has the symbol 'I'.

- Resistance: A material's opposition to an electrical current. Is measured in Ohms (Ω) and has the symbol 'R'.

- Power: The rate of doing work. Is measured in Watts and has the symbol 'P'.

- Energy: The capacity for work, the product of power and time. Has the symbol 'E'. The basic unit of energy is the Joule, but electrical energy is normally expressed in Watt hours (Wh) or kilo Watt hours (kWh). One kWh is 1000 Wh.

The relationship between these units is:

$P = VI$ *or* $V = P/I$ *or* $I = P/V$	Power equals voltage multiplied by current. This can also be expressed in the other two forms shown.
$V = IR$ *or* $I = V/R$ *or* $V = I/R$	Voltage equals current multiplied by resistance. Again there are two other forms shown. This is known as 'Ohm's law'.
$P = I^2R$	Power equals current squared multiplied by resistance.

2.2 Alternating and Direct Current

You will see references to Alternating (AC) and Direct Current. Direct current is where the current flows in one direction; alternating current reverses many times per second. Mains (grid) power is invariably AC as it can be transformed simply from one voltage to another, thus allowing it to be transmitted great distances. Batteries and photovoltaic modules produce DC.

Power calculations as detailed in section 2.1 are the same for AC and DC elements of the system.

2.3 The Photovoltaic Effect

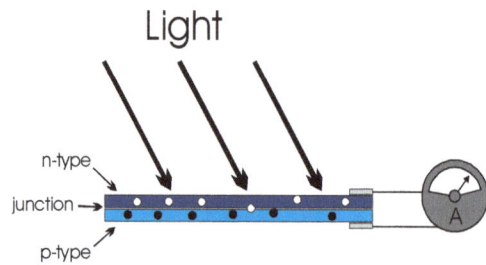

figure 1: the photovoltaic effect

The photovoltaic effect is the means by which solar panels or 'photovoltaic modules' generate electricity from light. A solar cell is made from a semiconductor material such as silicon. Impurities are added to this to create two layers, one of *n-type* material, which has too many electrons and one of *p-type* material which has two few. The junction between the two is known as a *p-n junction*. This process is known as *doping* and is the same technique used to manufacture transistors and integrated circuits (silicon chips).

Light consists of packets of energy called *photons*. When these photons hit the cell, they are either reflected, absorbed or pass straight through, depending on their wavelength. The energy from those which are absorbed is given to the electrons in the material which causes some of them to cross the p-n junction. If an electrical circuit is made between the two sides of the cell a current will flow. This current is proportional to the number of photons hitting the cell and therefore the light intensity.

2.4 Modules

A photovoltaic or PV module is commonly made from a number of cells connected together in series. This is because each cell only produces a voltage of about 0.5 Volts. It is usual in stand-alone applications for there to be 36 cells connected together to provide a voltage of about 18 – 20 Volts. This forms a module which can be used to charge a 12 Volt battery.

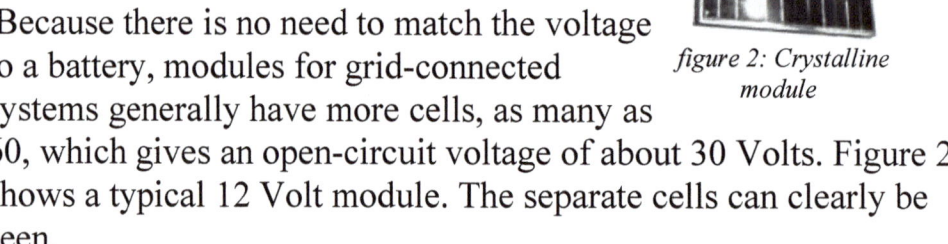

figure 2: Crystalline module

Because there is no need to match the voltage to a battery, modules for grid-connected systems generally have more cells, as many as 60, which gives an open-circuit voltage of about 30 Volts. Figure 2 shows a typical 12 Volt module. The separate cells can clearly be seen.

figure 3: thin-film module

There are also 'thin film' modules where the separate cells are formed as part of the manufacturing process. Figure 3 shows such a module. This technique is employed for the small solar panels which are fitted to calculators and similar devices. They are much cheaper to manufacture but deliver lower efficiency. This means

www.solar-power-answers.co.uk

that less of the light which hits them is converted to electricity. Recent advances in technology, however, have made larger and more efficient thin-film modules available.

Often a number of modules will be connected together into an *array* in order to provide more power than a single module can provide.

2.5 Energy Storage

Photovoltaic modules generate electricity only when there is light falling on them, and the amount of power generated is proportional to the light intensity. This means that a way has to be found of storing the electricity generated and releasing it when it is needed. The normal method is to use the surplus power to charge a lead-acid battery. This is the same type of battery as used in cars, although the different requirements mean that a car battery is not suitable, instead a *deep-cycle* battery is needed.

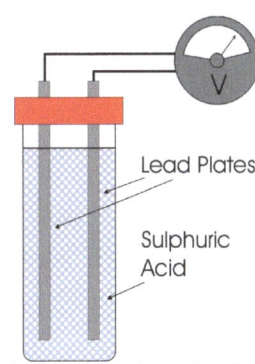

Lead Plates

Sulphuric Acid

figure 4: lead-acid cell

A battery is made up of a number of *cells*, each consisting of two lead plates in a container of dilute sulphuric acid. Each cell has a nominal voltage of 2 Volts, so a number are connected in series, for example 6 cells forms a 12 Volt battery.

Grid-connected systems do not need batteries. Effectively electricity is 'stored' by sending any surplus power to the grid, and using the grid to supply power when demand exceeds supply.

2.6 Control and Conversion

The electricity generated by the photovoltaic effect is low voltage direct current

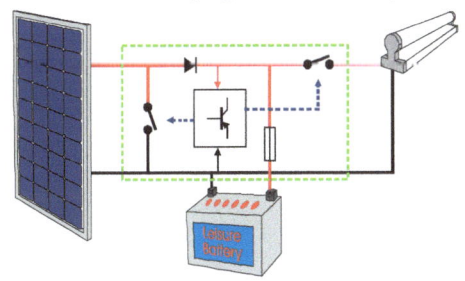

figure 5: controller operation

(DC) whereas mains electricity is much higher voltage alternating current (AC). This means that additional devices may be needed to control the battery charging process and convert the power to the correct voltage. The two most commonly used devices are the *photovoltaic controller* and the *inverter*. The controller makes sure that the battery is neither over-charged or over-discharged. The purpose of an inverter is to convert low voltage DC into higher voltage AC. It does this by first turning the DC power into AC and then using a transformer to step up to a higher voltage.

As grid-connected systems do not have batteries, there is no requirement for a controller. Instead the solar modules are connected to a '*synchronous inverter*', which synchronises the DC power from the solar array with the AC mains power in the house, and feeds any surplus to the grid, generally through a meter. The householder is usually paid for this surplus electricity, either at a different rate to the incoming power, or as at the same rate as a reduction to the bill *(net metering)*.

2.7 Operation

The principles of operation of a typical stand-alone solar power system are shown in figure 6. Electricity is generated in the form of low voltage DC by the photovoltaic modules whenever light falls on them.

This power is routed through a controller, which feeds whatever

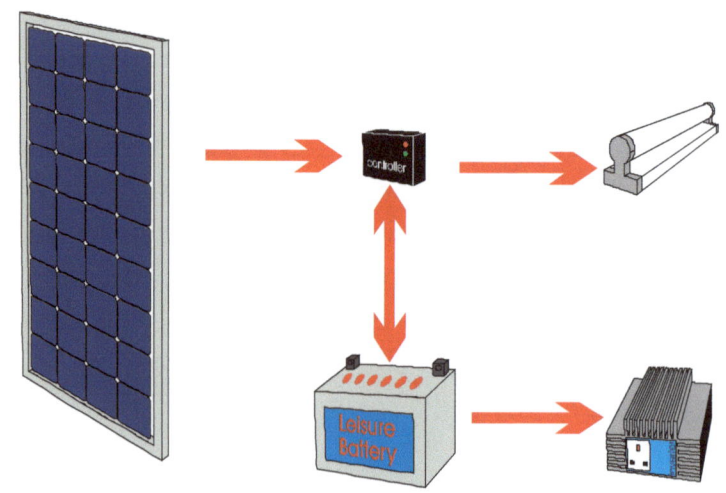

figure 6: power flow

power is necessary to any DC appliances such as lights and uses any surplus to charge a battery. When there is less power being generated than the appliances are using, power flows from the battery to the appliances. The controller monitors the battery state of charge and disconnects the appliances if the battery becomes very discharged.

Any AC (mains) appliances are connected to the inverter. This is not connected to the controller but directly to the battery. It incorporates its own control mechanism to ensure that the battery is not over-discharged.

3 Suitability

Before starting to design a solar power system it is important to assess whether solar power provides the best solution to the problem at hand. Solar power is best suited to applications where:

- The energy requirement is modest.

- There is no other source of power available.

- There is a good solar resource.

Despite this, there may be other good reasons for using solar power, for instance a concern for either the local or global environment, planning constraints or similar issues.

In the case of grid-connected systems, likely considerations are:

- Grid connection available

- Availability and rate of feed-in tariff

- Existing energy demand

- Good solar resource

For grid-connected systems, it is not necessary to fulfil all of the energy requirement. Therefore it is likely that system sizing will be as much a function of budget as of energy requirement. The sizing calculations can be followed to derive an estimate of the energy production and therefore the return on investment.

3.1 Energy requirement

The amount of energy which is required has a direct bearing on the size and cost of any proposed solar power system. The energy requirement can be reduced as discussed in a later chapter, however there are some applications for which solar generated electricity is very rarely suited. These include space heating, cooking, water heating and any other application where a large amount of heat is required. It may be possible to meet some of these requirements by more direct capture of solar energy, such as solar water heating systems or passive solar building design. These techniques are outside the scope of this manual, but see the Solar Power Answers website for more information.

There are some applications which easily lend themselves to solar power, such as lighting and computing, but most things will need to be assessed on a case-by-case basis.

3.2 Other power sources

One of the major factors affecting the choice of solar power is the availability of other potential sources of power. These may include such things as gas, diesel, kerosene and firewood. The most important however is mains electricity. If mains electricity is available it is very unlikely that a stand-alone solar power system will be economically viable except for very small energy requirements where the standing charge is likely to greatly outweigh the cost of the energy. A grid-connected system, of course, will offer a viable alternative.

The usefulness of any other source of power is determined by the nature of the energy form required. It isn't usually sensible to use electricity for heating, as heat is best obtained either directly by solar heating panels or by burning fuel, ideally wood from managed forests as this is a renewable resource. Lighting is almost certainly better delivered by solar or possibly wind power.

The reasons for choosing a certain fuel may be complex. For example, bottled gas may be a good fuel in a village close to a main road, however in a mountain village the cost of transport may make it impractical.

3.3 Solar resource

The availability of a good solar resource has a strong influence on the cost-effectiveness of a solar power system. A country in equatorial Africa offers great possibilities for solar power, not just because of the lack of other forms of power but also because of the high levels of sunshine throughout the year.

This does not mean, however, that solar power is impractical in countries further from the equator. In some remote parts of Great Britain, for example, the cost of connecting to mains electricity can be prohibitive. In this context solar power can be very competitive for moderate energy requirements.

Ultimately it may be impossible to decide whether or not solar power is suited to a particular application without following the design process. This way an estimate of the likely cost over the life of the project can be produced, which can then be compared with the costs of the alternatives. The capital costs of solar power systems tend to be high, however the running costs are low owing to the lack of any fuel costs and low regular maintenance requirements.

www.solar-power-answers.co.uk

4 System Components

In order to design a solar power system it is helpful to have a basic understanding of the various system components and their operation. The following paragraphs describe those components which will commonly be encountered.

4.1 Modules

4.1.1 Types

As already discussed there are two basic types of solar module, crystalline and thin-film. The characteristics of these are similar but the method of manufacture is very different.

4.1.1.1 Crystalline

A crystalline module is made from a number of discrete cells, usually 36 for a 12 Volt module. These cells have to be assembled and soldered together by hand, which goes some way to explaining the relatively high price of crystalline modules. Each cell is made from a wafer, composed either of a single crystal (monocrystalline) or many crystals (polycrystalline) of a semiconductor material, usually silicon. The monocrystalline method produces cells of slightly higher efficiency, but for all practical purposes they can be regarded as the same. Polycrystalline modules can be distinguished by the obvious crystalline appearance of the cells.

4.1.1.2 Thin film

Thin film or "amorphous" modules are made by a different process. A thin film of semiconductor material is deposited on a substrate, usually glass. This substrate forms the body of the module. A laser is then used to score the material in order to produce individual cells, which produces the characteristic

figure 7: thin-film modules

striped appearance. This method uses less energy, less of the semiconductor material and is easier to automate. The modules thus produced are therefore lower cost. Currently, however, commercially available thin film modules display significantly lower efficiencies than crystalline modules. This limits their use to applications where there is no size restriction on the array and adds to the cost of installation as more modules are necessary.

4.1.2 Operation

Figure 8 shows the relationship between voltage and current for an imaginary 12 Volt 36 cell module, the most common configuration. The two curves represent different insolation levels.

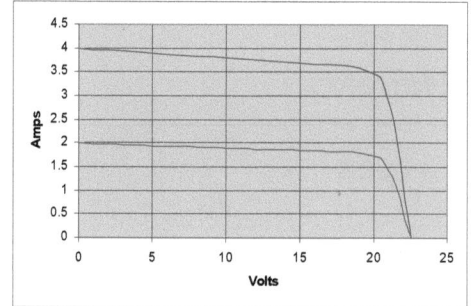

The current that a photovoltaic module will deliver into a short circuit is known as the short-circuit current or I_{sc}. This is proportional to the insolation, so the more the sun shines the greater the current.

The open circuit voltage or V_{oc} is determined by the number of cells in series, and is not significantly affected by the insolation. You

figure 8: Power characteristics

can see that over the working voltage of a 12 Volt load such as a battery the current is nearly constant for a given value of insolation.

4.2 Batteries

4.2.1 Types

There are many different battery technologies available today. However it is one of the oldest, the Lead-Acid battery, which is most suited to stationary solar power applications. There are two main reasons for this; a large amount of energy storage costs very little compared to other technologies and it operates over a narrow voltage range which makes it ideal for powering common appliances. This type of battery does have its disadvantages, notably the fact that it is easily damaged by excessive discharge. Each *cell* of a lead-acid battery has a nominal voltage of 2 Volts, hence a 12 Volt battery is constructed of 6 cells in series. Lead acid batteries are usually available as 2 Volt *cells* or 6 Volt or 12 Volt *monoblocs*, i.e. a number of cells combined to make a battery. A standard car battery is an example of a 12 Volt monobloc.

4.2.1.1 Starting batteries

Starting batteries such as car batteries are easily available at very low cost. They are designed to deliver a very large current for a short time. Contrary to common belief this does not result in a heavy discharge, usually no more than 5% of the battery's total capacity.

The demands of solar power systems require that the batteries are frequently discharged by 50% or more, and thus a starting battery is unsuitable. Attempts to use starting batteries in this way results in a very short life and is a false economy.

www.solar-power-answers.co.uk

4.2.1.2 Deep-cycle batteries

The term 'deep-cycle' refers to batteries that are designed for regular discharging by 50% or more. The term is applied to many different forms of battery from small 6 or 12 Volt batteries to much larger batteries consisting of separate 2 Volt cells. Most traction batteries, that is those designed to propel electric vehicles such as fork-lift trucks, can also be considered to be deep-cycle. The majority of deep-cycle batteries have a liquid electrolyte (acid) which is vented to the atmosphere. Sealed types with the electrolyte in the form of a gel are also available, although their higher cost limits their use.

4.2.1.3 Leisure batteries

The term 'leisure battery' refers to a battery which is a compromise between the low cost of a car battery and the long life of a true deep-cycle battery. They have a much longer life than a car battery when regularly discharged and are much less expensive than a true deep-cycle battery. Their use is common in applications such as caravans, where the usage pattern is not as intensive.

4.2.2 Operation

4.2.2.1 Charging

The voltage at which a lead-acid battery is charged must be strictly regulated. If the charging voltage is too high, then excessive gassing will occur, leading to loss of electrolyte and possible plate damage. On the other hand, too low a voltage will lead to the plates becoming 'sulphated' which causes a loss of capacity. Figure 9 shows the relationship between voltage and current in a constant voltage charging regime.

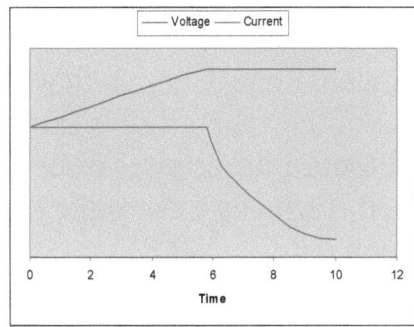

figure 9: charging

www.solar-power-answers.co.uk

4.2.2.2 *Discharging*

Batteries must be protected from damage by over-discharge. As the battery discharges the voltage at the terminals decreases. Figure 10 shows the terminal voltage of a lead-acid battery at differing rates of discharge. You can see from this how it is impossible to deduce the state of charge from the battery voltage alone, and therefore why some kind of over-discharge protection is necessary.

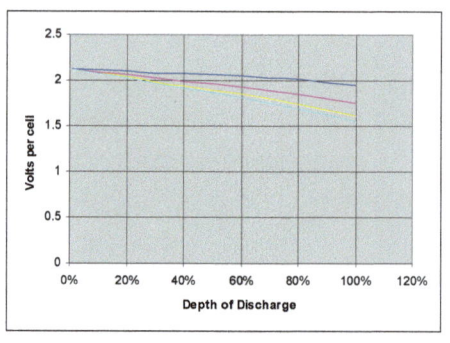

Figure 10: Battery discharge

4.3 Controllers

4.3.1 Function

The primary function of the controller is to regulate the charging of the battery. Many controllers also perform a number of secondary functions, the most common of which are the protection of the battery from over-discharge (low-voltage disconnect) and status monitoring by means of lamps or LCD displays.

4.3.2 Operation

A standard controller can function in two basics ways; *shunt regulation*, where the output of the solar array is shorted to control its output and *series regulation*, where the output of the array is disconnected by some form of switch. The switching method is usually electronic, although electro-mechanical systems may be encountered.

There is a third type of controller, the *Maximum Power Point Tracker (MPPT)*. MPPT controllers operate by maintaining the input (solar) voltage at the point where the modules are generating maximum power, i.e. Voltage x Current. This voltage is higher than the battery voltage, so they use a switch-mode circuit to convert the output voltage to the correct battery charging voltage. This results in greater usable power from the same array, or a smaller array to provide the same power. MPPT controllers are more expensive, and therefore cost-effective in larger systems.

4.4 Inverters

4.4.1 Function

The function of an inverter is to transform the low voltage DC of a lead-acid battery into higher voltage AC which may be used to power standard 'mains' appliances. An inverter is necessary where appropriate low voltage appliances are unavailable or expensive or in larger systems where it is necessary to distribute the power over a wide area.

4.4.2 Synchronous Inverters

A *synchronous inverter* is used in a grid-connected system to connect the solar array to the grid. It synchronises its output to the frequency and voltage of the incoming mains (grid) power, and incorporates safety circuitry which disconnects the output if the mains is disconnected. This feature is to ensure that anyone working on the line will be safe, but means that a synchronous inverter is useless without mains power. So, a grid-connected array is unable to operate as a backup power supply in case of loss of mains.

4.4.3 Operation

For our purposes there are two types of stand-alone inverter; *sine wave*, which closely mimics the waveform of mains electricity and *modified sine wave*, which is more accurately described as a square-edged waveform with similar characteristics to a sine wave. Figure 11 shows the two waveforms.

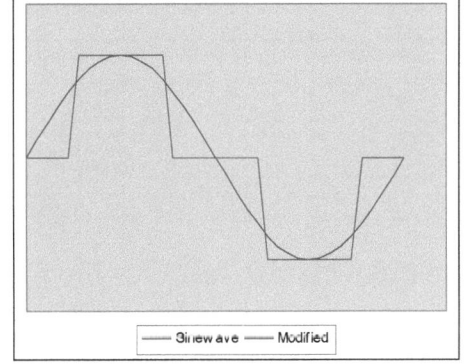

There are advantages and disadvantages to both types. Generally speaking cheaper small inverters will be of the modified sine wave type. However certain equipment may not operate correctly or may be noisy. These problems will not occur with a sine wave

figure 11: inverter waveforms

inverter, as the waveform is identical to that delivered by the mains. The efficiency of recent models of sine wave inverter is comparable to that of modified sine-wave units.

www.solar-power-answers.co.uk

5 Design

5.1 The design process

The system design process consists of four major steps. These are:

- Initial estimates
- Site survey
- System sizing
- Component selection

The order in which these are performed will depend on the amount of information available in advance and factors such as the ease of visiting the site. If detailed information regarding the location of the site and the intended loads are available then it may be possible to size the system before a site visit takes place.

5.2 Initial estimates

Before the commencement of the design process proper, you will need to have at least a rough idea of what you hope to achieve, for example: *"To provide lighting and refrigeration for a holiday home"*. From this it will be possible to produce initial estimates to feed into the system design process. The following paragraphs expand on this example.

5.2.1 Load estimates

In order to estimate the load requirement we need to get an idea of the type of usage the system will be put to. For the above example of a holiday home we should be able to discover how many rooms it has and how many people will be likely to be in residence. If we make the following assumptions:

- There are 3 rooms, one of which is a bedroom, therefore;
- There will be no more than 2 people in residence.

Then we can estimate the lighting and refrigeration as follows.

5.2.1.1 Lighting

From the above we know that there are three rooms, so the total number of lights required is 3. Now we need to estimate the average daily usage of each light.

The first thing we can deduce is that, if there are two people then there need not be more than 2 lights on at any one time. Then we can make an estimate of the amount of time between darkness falling and the residents retiring. Let us say that is 8 hours.

Now, let us assume that the occupants spend half of this time together. In that case one of the lamps will be on for half of the time (4 hours) and the other for all of the time (8 hours). So this gives us a figure of 3 lamps and a total of 12 hours, hence each lamp is on for an average of 4 hours per day.

Lastly, you need to estimate the power consumption of each lamp. This is a matter of picking a type of lamp which you think will be suitable by examining the lamps available to you. For this example let's assume that an 11 Watt, 12 Volt fluorescent lamp is selected.

5.2.1.2 Refrigeration

Estimating the refrigeration requirement is rather more straightforward. All that is necessary is to find a suitable 12 Volt compressor refrigerator in a manufacturer's catalogue and look up its daily energy consumption. This will be determined by the ambient temperature so an estimate of that will be helpful. For the purposes of this example let's assume an energy consumption of 600 Wh/day at a 25°C average.

5.2.1.3 Other loads

The energy requirement for any other loads is calculated in the same way as that for the lighting. The power consumption of each item is multiplied by the number of hours it will be used in a day to give the energy consumption in Wh/day.

5.2.1.4 Phantom loads

Phantom loads is the name given to those appliances which use power even when they are switched off. Example include audio-visual equipment such as televisions, DVD players and anything which has the power supply built into the plug or cable such as laptop computers and mobile telephones. Anything that falls into this category should be unplugged when not in use or provision made to switch off its supply. However there may be appliances which need to remain plugged in. The standby consumption of any such appliance must be treated as an additional load which is in use for all the hours of the day that the appliance itself is not in use.

5.2.2 Location

The intended location of the system will determine the solar resource which is available. This in turn will allow the size of the solar array to be calculated. For the purpose of this example, let's assume that the holiday home is in northern Portugal.

www.solar-power-answers.co.uk

5.2.3 First iteration

From the initial problem:

"To provide lighting and refrigeration for a holiday home"

we have now arrived at:

"To design a solar power system to be installed in northern Portugal, to power three 11 Watt lamps for an average of 4 hours per day and a refrigerator with an energy requirement of 600 Watt-hours per day"

Following the system sizing process (section 5.4) will show whether this is a practical system. If not, then make changes to the requirements and start again. For instance, in this example the refrigerator consumes far more energy than the lighting. If the system is likely to be too expensive, then consider using a gas refrigerator instead. The capital cost will be lower, but there will be a fuel cost to take into account.

5.3 Site Survey

In some cases it may be necessary to complete the design without having visited the site, in which case certain assumptions will need to be made. In these circumstances it would be advantageous to obtain photographs of the site and the surrounding area if possible, or failing that a detailed description.

The various points of the site survey are covered in the following paragraphs. It will be helpful to take photographs of the site for reference later. Pay particular attention to those areas chosen for the various system components; as the design progresses these will be invaluable.

5.3.1 Shading

The first and most obvious check is to ensure that the sun actually shines on the site. From the projected position of the system survey the horizon over the entire arc of the sun. In the northern hemisphere you should be looking towards the south, east and west and in the southern hemisphere the north, east and west. Very close to the equator the sun passes virtually overhead, so only the east and west are important.

You should be looking for anything which will shade the solar array at any time of the year, including such things as:

- Trees. If it is winter when you visit, remember that some trees will look very different in summer. Also include sufficient space for 20 years of growth.

- Hedges. Again allow for these to grow significantly during the life of the system.

- Mountains and hills. Remember that the sun will be much closer to the horizon in the winter. If it is summer when you visit, ask someone local where the sun rises and sets in the winter.

- Buildings. Ask around to ensure that no building work is planned which will obscure the site.

- Climate. Find out if there is anything unusual about the climate in the local area such as sea mist.

Try to imagine what the site will look like all the year round and in years to come. You may find it helpful to make a sketch of the surrounding area for later reference.

5.3.2 Array location

It will be necessary to find a position for mounting the solar panel or array. If the system is to be installed in a building, then it is common for the array to mounted on the roof of the building as described below. If this is not possible then an alternative site will need to be found.

5.3.2.1 Roof

The ideal is for a roof with a slope towards the south if in the northern hemisphere or the north if in the southern hemisphere. The angle of this slope to the horizontal needs to be about equivalent to the angle of latitude plus 15°. It is very unlikely that these conditions will be met, however the roof is still likely to be the best place if it slopes in roughly the right direction or is flat. If it is flat, however, it will be necessary to arrange some type of angled support such as that used for ground mounting.

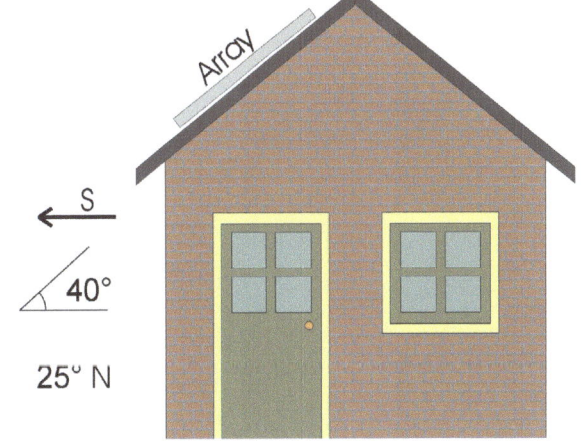

figure 12: roof mounting

If possible gain access to the roof in order to survey it more thoroughly.

figure 13: solar roof

Check the following:

- Shading. See section 5.3.1.

- Direction. Use a compass to check what direction the roof slopes towards.

- Angle of slope. Use a spirit level to measure the angle of the roof from the horizontal.

- Material of construction. If necessary also check underneath the roof to see what fixings will be needed and ensure that the structure is strong enough to support the weight of the array.

- Area. Measure and record the dimensions of the usable part of the surface of the roof. Estimate whether this will be sufficient for the size of array that is likely to be needed.

If it appears that the roof will not be suitable then it will be necessary to find a site for an alternative form of support.

5.3.2.2 Ground mounting

In the absence of a roof or similar structure to mount the array on it will be necessary to use some form of support structure. Solar equipment suppliers sell different types of structure or it may be possible to fabricate a support on site. There are two basic types as illustrated in figure 14:

- Ground mounted, where the structure is a frame mounted on the ground which requires a foundation, and

- Pole mounted, which can be attached to an existing pole or a pole erected for the purpose.

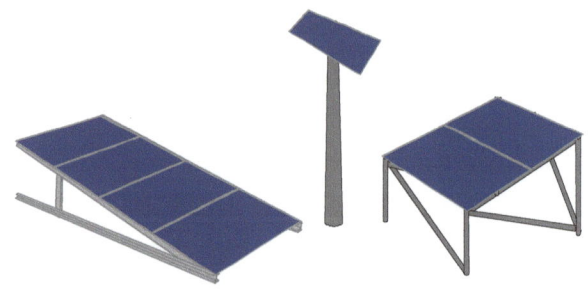

figure 14: support structures

The survey should take account of:

- Shading.

- Ground conditions, for the purpose of building foundations.

- Available area of ground.

- Distance from location of batteries for cable sizing.

- Any suitable poles.

figure 15: ground mounted array

5.3.2.3 Other options

There may be other mounting systems available. For example, figure 16 shows a system where the modules are mounted with other system components on a south-facing gable end. If there is no potential for roof or ground mounting then it may be that there is another solution which will suit the needs of the planned installation.

figure 16: gable end

5.3.3 Batteries

5.3.3.1 Location

A suitable position must be found for the batteries. This may be a room within a building, a separate building or a place where some kind of housing can be erected. The following conditions need to be met:

- Environmental protection. The batteries need to protected from rainfall, direct sunlight and extremes of temperature.

- Ventilation. All lead-acid batteries, even sealed types, need to be adequately ventilated in case of gassing.

- Protection from sources of ignition. When under charge vented lead-acid batteries give off an explosive mixture of hydrogen and oxygen, so must not be exposed to any sources of ignition such as naked flames.

- Personal safety. Because of the explosive gasses given off and the potential for extremely high currents, batteries must be kept in a secure place away from children.

Consideration should also be given to the likely location of the other system components. The aim should be to ensure that the cable runs are kept as short as possible. This is most important when considering the cable between the battery and controller.

5.3.3.2 *Mounting*

Where a suitable location indoors has been identified, it will often be acceptable to place the batteries directly on the floor. If not, it may be necessary to mount the batteries in a battery box or on racking in order to make best use of the available space or to offer them adequate protection.

Consideration should be given to the likely shape of any such boxes or racking, and the area measured to ensure that the batteries will fit as intended.

5.3.4 Control equipment

It is usual for the controller, inverter and other control equipment to be wall-mounted. An indoor location will be needed, as close to the batteries as possible. There is often a restriction on how long the battery cables can be, so it is important to ensure that this can be met.

Inverters in particular are often quite heavy. Assess the chosen wall so as to ensure that it will be able to take the likely weight of the equipment.

5.3.5 Loads

The site survey provides an opportunity to more accurately assess the loads, for instance the lighting requirement, where the system is to be installed in an existing building. If possible ask people about the use to which each room is put and times during which it is occupied.

It is possible that there may be existing electrical wiring, for example if a generator is used. It may be possible to use all or part of this for the solar application. If this is the intention, then inspect the wiring and record its configuration.

5.3.6 Cabling

Take the opportunity to consider likely routes for the cabling, especially the heavy cables running from the controller to the array and the batteries. Measure the approximate length of these cables so that they can be correctly sized.

www.solar-power-answers.co.uk

5.4 System sizing

System sizing is the process of determining the size of the various system components, for instance the peak power rating of the array or the current rating of the controller. The selection of the components themselves is covered in section 5.5.

I have written a Microsoft Excel template, 'Solar sizing.xlt', which performs the calculations described in this chapter, which is available for download from the Solar Power Answers website. If this is not available for whatever reason then the calculations can be performed by hand or with a calculator.

There are a number of steps to be followed. It may also be necessary to perform a number of iterations; if the result of the sizing process is not as expected it may be necessary to repeat the process a number of times.

5.4.1 Loads

The first part of the process is to calculate the daily energy requirement of the proposed system in Watt-hours per day.

Appliance	Quantity	Rating (W)	Usage (h)	Wh/day each	Total Wh/day
Lamp	3	11.00	4.0	44	132
Computer				0	0
Television				0	0
Radio				0	0
Fridge	1			600	600
Other				0	0
				0	0
				0	0
				0	0
				0	0
				0	0
				0	0
				0	0
Total					732

figure 17: load calculation

5.4.1.1 Assessment

For each load determine the power rating in Watts. This may be found on the appliance or in the manufacturer's data. The power rating may also be stated in kilowatts or kW where 1 kW = 1000 W. If this information is not available then appendix 5 gives approximate power ratings for common appliances. Now determine now many of each appliance is needed and the average daily hours of use. Enter these figures in the appropriate columns as shown in figure 17 or use this equation:

$E = n \times P \times T$

Where:

E is the energy requirement in Wh/day
n is the number of appliances
P is the power rating in Watts
T is the average usage in hours

For any mains voltage or AC appliances it is necessary to account for inverter efficiency. This is because some power is lost when an inverter converts low voltage DC into high voltage AC. Divide the result above by 0.9 (90%) unless you know the efficiency of the actual inverter that will be used.

Now total the results for all the loads:

$$E_T = E_1 + E_2 + \ldots$$

Where:

E_T is the total energy requirement

$E_1, E_2, \ldots$ are the energy requirements of the individual loads

This will give a total figure for the energy requirement of the system.

For the example above the result is 732 Wh/day.

5.4.1.2 Optimisation

It is worth paying some attention to optimising the load, that is decreasing the energy consumption to its practical minimum. From the figures above it will be apparent which of the loads are the most significant in terms of energy consumption. Consider whether there are any gains to be made, for instance by using a smaller number or more efficient appliances, using 12 Volt instead of 230 Volt appliances or using a different energy source for some appliances.

Repeating the steps at sections 5.4.1.1 and 5.4.1.2 until the optimum is reached will pay dividends later in the process.

5.4.1.3 System voltage

At this stage it will help to decide on the system voltage, that is the voltage of the battery bank. The choice is normally dependent on the loads which it is necessary to power. If there are loads which are 12 Volt, then obviously it makes sense for the system voltage to be 12 Volts. However if there is a large 230 Volt requirement then it may be necessary to consider 24 Volts or even 48 Volts in order to obtain a suitable inverter.

There are no fixed rules for the choice of system voltage. On balance it is probably best to use 12 Volts unless there are compelling reasons to use a different voltage. This choice may also affect your choice of loads, and it may therefore be necessary to repeat the assessment and optimisation processes.

5.4.2 Solar array

The size of the solar array is determined by the daily energy requirement and the solar resource or *insolation* available to the system. The greater the energy requirement the larger the solar array needs to be and the greater the insolation the smaller the array.

The sizing of the solar array for a grid-connected system uses the same principles, although it will not necessarily be sized to match the power consumption in the same way, as it may only fulfil part of the requirement.

www.solar-power-answers.co.uk

5.4.2.1 Insolation

Insolation is a measure of the amount of solar energy falling on an area. The usual measure is *kWh/m²/day*. That is kilowatt-hours (thousands of Wh) per square metre per day.

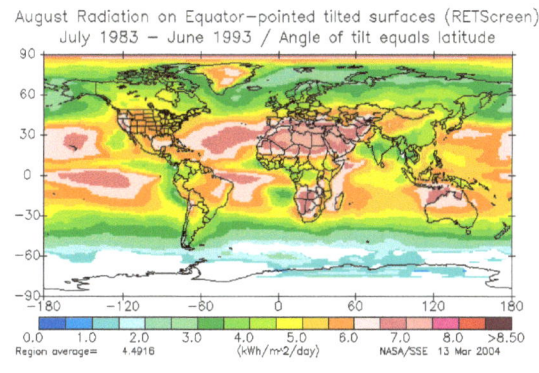

figure 18: insolation map

Insolation data may be obtained from a variety of different sources such as meteorological agencies. If a local source of data is not available then the data from NASA's web site at http://eosweb.larc.nasa.gov/sse/ is an excellent resource which is free at the time of writing.

If data is not available then appendix 1 provides a set of global insolation maps derived from the NASA data which will provide sufficient accuracy for most purposes.

The aim is to determine a figure for 'design insolation'. This is the minimum daily average insolation available to the system. The figure used should be from the month with the least insolation, based on whichever months of the year the system is intended to be used. The figure arrived at is likely to be between 1 and 6 kWh/m²/day.

For our holiday home example, assuming that it may be used at any time during the year, a figure of 3.5 kWh/m²/day is appropriate. This represents the lowest monthly average insolation for Portugal, from the insolation maps in appendix 1. If the building were only occupied at certain times of year, then the lowest figure from the months it is occupied could be used, which would likely be a larger number and therefore reduce system costs.

5.4.2.2 Efficiency

Having determined a figure for design insolation the efficiency of the battery charging process must be considered. There are two factors to take into account; power point efficiency and charge cycle efficiency.

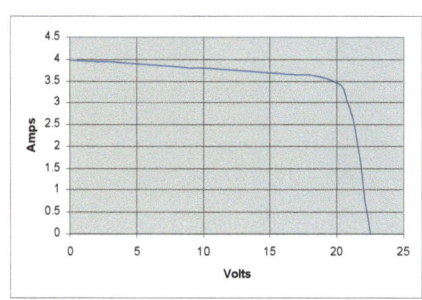

figure 19: typical output

25

As shown in figure 19, the peak power output of a solar module is produced at the 'knee' of the output curve. In this example this is at 20 Volts, where the current is 3.5 Amps. Therefore the peak power output is:

20 x 3.5 = 70 Watts

However the battery charging voltage is likely to be between 13 and 14 Volts. From the graph we can see that at 14 Volts the current is approximately 3.75 Amps. This gives a power output of:

14 x 3.75 = 52.5 Watts

Hence the efficiency is:

52.5 / 70 = 0.75 or 75%

If you have a copy of the output curve for the particular modules which you intend to use, then calculate the efficiency as described. Otherwise use the figure of 0.75 as this is a reasonable approximation for the purposes of system sizing.

In the case of an MPPT charge controller, the conversion efficiency of the controller should be used. This figure should be quoted by the manufacturer, and will be greater, typically around 95%.

5.4.2.2.2   Charge cycle efficiency

The charge cycle efficiency is a measure of the proportion of the energy used to charge a battery which is returned when the battery is discharged. The actual efficiency of a particular battery may be obtained from the manufacturer, however an approximation will suffice. For this purpose assume an efficiency of 0.95 or 95%.

5.4.2.3 Sizing calculation

All the variables necessary to size the solar array are now known. Enter the values into the spreadsheet as shown in figure 20, or proceed as follows:

Insolation	3.5	kWh/m^2/day
Charge cycle	95%	% efficiency
Power point	75%	% mismatch
Load	732	Wh/day
Holdover	3	Days
System Voltage	12	Volts
Depth of discharge	50%	% d.o.d.

PV requirement	294	Wp
Battery requirement	366	Ah

figure 20: sizing calculation

> The sizing calculation is:
>
> S = (E / i) / (e$_{pv}$ x e$_{bat}$)
>
> Where:
>
> *S = Array size in peak Watts or W$_p$*
> *E = Daily energy requirement in Wh/day from section 5.4.1*
> *i = Insolation in kWh/m^2/day*
> *e$_{pv}$ = Power point efficiency*
> *e$_{bat}$ = Charge cycle efficiency*
>
> It can be seen that the result of this calculation is not dependent on the system voltage, as it refers to the power output of the array rather than the current or voltage.

Figure 20 gives the result of this calculation for the example holiday home system.

5.4.3 Battery

Battery sizing is the process of ensuring that there is sufficient battery capacity to support the loads during such times as there is insufficient energy available from the solar array. As battery capacity is relatively cheap, it may be thought that it is impossible to have too great a battery capacity. This is a fallacy, as it is important to ensure that the battery is fully charged on a regular basis to prevent damage through sulphation. An overly large battery in comparison to the size of the array will not reach full charge as it will require a greater charging current than the array can deliver.

Battery capacity is measured in Ampere-hours (Ah) at the system voltage, and is derived as a function of the daily energy requirement, the 'holdover' requirement and the 'depth of discharge' limit.

5.4.3.1 Holdover

The holdover is simply defined as the number of days that the load is required to operate without any charging input from the solar array. It should be noted, however, that at most latitudes there is no such thing as a day with no sun. Even overcast winter days can provide some useful charging input, so no system will ever be required to operate entirely from the battery for the holdover period.

The required holdover period is determined by the security of supply required. There is no hard and fast rule, but for general applications such as lighting and

27

domestic purposes a figure of 3 days should be adequate. For more critical application such as medical refrigeration a period of 7 days should be considered. If there is another source of charging input such as a diesel or wind generator then it is possible to reduce the holdover. If a modular battery system is chosen then it will be relatively easy to increase the battery capacity at a later date should it prove necessary.

5.4.3.2 Depth of discharge

The *depth of discharge* is the proportion of the battery's capacity that can be used by the loads without recharging. It is the opposite of the *state of charge*; a depth of discharge of 80% is equivalent to a state of charge of 20%.

The design depth of discharge is determined by the type of battery used and the expected life, balanced against the cost of the battery. A leisure type battery will typically be used to a depth of discharge of between 30% and 50%, whereas a deep-cycle or traction battery will be discharged to between 50% and 80%. This should be considered a practical maximum, as there is no type of lead-acid battery which can safely be completely discharged on a regular basis.

It is a myth that lead-acid batteries last longer when regularly deeply discharged. The life of a lead-acid battery is shortened by each charge / discharge cycle, by an amount roughly proportional to the depth of discharge of that cycle.

5.4.3.3 Sizing calculation

Once you have arrived at figures for the holdover and depth of discharge, enter these into the spreadsheet (figure 20). Alternatively follow this method:

The sizing calculation is:

$C = (E \times h / d) / V$

Where:

C = Battery capacity in ampere-hours (Ah)
E = Daily energy requirement in Wh/day from section 5.4.1
h = Holdover in days
d = Depth of discharge expressed as a decimal
V = System voltage

See figure 20 for the results of the battery sizing calculation for our example.

5.4.4 Allowing for expansion

Depending on the use that the system is going to be put to, you may need to consider the possibility of future expansion. In this case, the controller in particular should be sized to meet the future maximum size of the array. Batteries and modules can be added to, although it's always best to use only components that are the same as the ones already installed.

In the case of a later expansion of battery capacity, it is important to consider whether the existing battery is close to the end of its life. If it is, then it would be better to replace the entire battery with one of larger capacity, rather than add to an old battery, only to have to replace the existing cells a short time later.

5.4.5 Hybrid systems

It is common to combine solar power with other forms of generation, either renewable or conventional. Any supplemental battery charging source must be connected directly to the battery terminals and not via the solar charge regulator. The most common types of hybrid systems are covered briefly here.

5.4.5.1 Wind turbines

Wind turbines offer a good complement to solar photovoltaics. After all, there aren't many days which are neither sunny nor windy. In order to calculate the solar component of such a system it is necessary to know the monthly output of the wind turbine. Once this is known, then the solar part of the system can be sized as above, performing a separate calculation for each month of the year, and subtracting the daily contribution from the wind turbine (monthly output divided by number of days in the month) from the load requirement. You will also need to use the insolation figure for that particular month.

5.4.5.2 Diesel generators

In many systems a diesel or petrol engined generator is used either to ensure security of supply or to supplement the solar output during the winter months. The generator is usually wired through a changeover relay to replace the inverter when running, and also to a battery charger so that the batteries will be replenished at the same time.

The battery charger chosen should be a model designed for this type of application, and sized in consultation with the generator manufacturers. The maximum sized battery charger that a given generator can operate will be significantly smaller than the generator rating suggests.

Some inverters are capable of remotely starting a generator when the battery needs charging, thus allowing the system to be completely automated. Some incorporate a battery charger with automatic switching between inverter and generator. For a larger system with a backup generator these are a good choice.

5.5 Component selection

The selection of the components of a solar power system is determined by their electrical characteristics. However there are other factors including price, availability and the necessity for any parts to fit in the space available.

5.5.1 Solar array

The solar array consists of more than just the modules. There is also the support structure and cabling to consider. The selection of the solar array depends on the particular installation as follows.

5.5.1.1 Modules

The prime consideration when choosing modules is usually their cost per Watt. This is the price of the module divided by the peak wattage, for example if a 50 Wp module costs $200, then it is said to cost $4 per watt. This is not the only consideration however. The following points should also be taken into account:

- Physical size. It is likely that thin-film modules will be bigger than crystalline for the same power rating. Likewise a large number of small modules is likely to occupy a larger area than a small number of large modules. The space available for mounting may therefore determine which modules are chosen.

- Support structure. If it is intended to purchase a ready-made support structure then they may only be available for certain combinations of modules. Also a support structure for many small modules is likely to cost more than one for a few larger modules. The greater flexibility may, however, outweigh this disadvantage.

- Cabling and installation. Again there will be more cabling for a larger number of modules. This will increase the time needed for installation and the quantity of cable required.

- Fit to system. How closely it is possible to meet the system requirements should be considered. For instance, if a minimum of 60 Watts at 12 Volts is needed, then three 20 Watt modules would be a better fit than two 50 Watt modules, and may be cheaper. However, if the system were 24 Volts, then four 20 Watt modules would be needed, which would be more expensive.

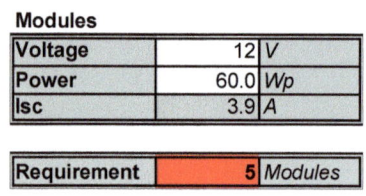

Modules		
Voltage	12	V
Power	60.0	Wp
Isc	3.9	A

Requirement	5	Modules

figure 21: module selection

Figure 21 shows the module selection function of the sizing spreadsheet. Enter the voltage and peak power rating of the chosen modules and the number of modules required is calculated. This can easily be calculated manually if the spreadsheet is not available. By entering the details of all the available modules and multiplying

www.solar-power-answers.co.uk

the result by the cost per module the optimum modules can be selected. Remember to check that the selected modules will fit in the space earmarked for them.

5.5.1.2 Support structure

The selection of a support structure is dependent on the results of the site survey. If it is intended to purchase a structure then this should be considered as part of the process of selecting the modules. The angle of tilt should be able to be set at the correct angle for the system. The correct angle from the horizontal is normally equal to the angle of latitude at the location where the system is to be installed.

It is perfectly feasible to manufacture a support system on site from steel or aluminium or even wood. Galvanised perforated steel angle is ideal. The design of the structure must take account of the worst case wind loading, which will be significant, especially if the structure is roof-mounted.

5.5.1.3 Cabling

It may be possible to purchase ready-made cables known as array interconnects. These are short cables cut to the correct lengths to connect the modules together and are resistant to ultra-violet light. Alternatively it is possible to make these on site; this is covered in the section on installation.

Grid-connected systems are usually connected with ready-made cabling with connectors included. This is because the system voltages are much higher, and therefore present a greater danger.

5.5.2 Battery

There are many options for the system battery, and for systems with a battery requirement of much more than a few hundred Amp-hours it is best to seek the advice of a specialist battery supplier. Batteries are available as individual cells or as 'monobloc', that is a number of cells in a single battery in the same way as a car battery.

5.5.2.1 Configuration

A battery can be made up in many ways. For example a 24 Volt, 200 Amp-hour battery may be configured as, for example:

1. Four 12 V, 100 Ah monoblocs in series / parallel.

2. Four 6 V, 200 Ah monoblocs in series.

3. Twelve 200 Ah cells in series.

Batteries		
Voltage	12	V
Capacity	100	Ah

Requirement	4	Batteries

figure 22: battery selection

For low cost domestic applications the normal choice is leisure batteries or small monobloc traction batteries. These are usually 12 Volt or 6 Volt monoblocs with a capacity between about 60 Ah and 110 Ah.

Remember that, as the battery capacity is expressed in Amp-hours, twice the number of cells will be required for the same capacity at 24 Volts than at 12 Volts. For example, a 12 V, 100 Ah battery may consist of a single monobloc. Two of these monoblocs in series would constitute a 24 V, 100 Ah battery; two in parallel a 12 V, 200 Ah battery.

Figure 24 shows the battery selection table of the sizing spreadsheet.

5.5.2.2 Lifetime

The life of a battery is normally expressed in two ways. 'Life in float service' is the life of a battery in years if it is always on charge and never discharged. This can be seen as the maximum life. 'Cycle life' is expressed as a number of cycles to a

particular depth of discharge, e.g. 300 cycles to 80% d.o.d. This is sometimes available from the manufacturer in the form of a graph or table showing the cycle life versus the depth of discharge such as that in figure 23.

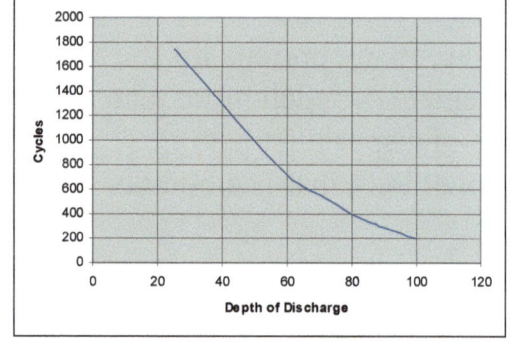

Determining the life of a battery in a solar power system is not straightforward. It is difficult to accurately predict the number or depth of discharges as this is determined by

figure 23: cycle life

the weather and the usage pattern. An approximation can be made by dividing the number of days in the year (365) by the number of days holdover in the system, so a system with 3 days holdover would perform approximately 120 cycles annually.

From this information can be determined the expected life of the battery, for example a battery with a cycle life of 1000 cycles to 50% d.o.d., in a system designed for 3 days holdover to 50% d.o.d., will last around 8 to 8½ years. The optimum battery life is reached when the cycle life is equal to the life in float service. Any further increase in capacity will not extend the life of the battery.

5.5.2.3 Sealed or vented

Traditional lead acid batteries usually include vented caps which allow lost electrolyte to be replenished. Many modern batteries are sealed. Sealed batteries come in three types; liquid, AGM and Gel. Many of the batteries which are generally referred to as 'gel' batteries are in fact AGM batteries, where the electrolyte (acid) is contained in an absorbent material between the plates, preventing spillage. In a true gel battery the electrolyte is a jelly. As a general principle AGM batteries can provide a higher current where gel batteries have a longer cycle life, although this is by no means universal.

In general, it is better to use a traditional vented battery wherever possible. Reasons for using a sealed battery include:

- Freedom from explosive gasses

- Maintenance free

- Ease of transporting

However these are outweighed in many applications by the significantly higher cost and frequently shorter lifespan, especially at high ambient temperatures.

5.5.3 Controller

The selection of the controller is determined by four factors; the system voltage, the array (input) current, the load (output) current and the type of battery. If it is not possible to find a controller with the correct specifications, then it may be necessary to change the system voltage and repeat the sizing calculations.

5.5.3.1 Voltage

The chosen controller must be able to operate at the system voltage chosen in the section on sizing. Many controllers can operate at more than one voltage; some automatically select the correct voltage.

5.5.3.2 Array current

The maximum array current is the short circuit current (I_{sc}) of an individual module multiplied by the number of modules in parallel. For example, in a 12 Volt system with four modules with a short-circuit current of 4 Amps each, the array current is 16 Amps. For a 24 Volt system with the same modules, there are only two in parallel so the array current is 8 Amps.

5.5.3.3 Load current

If the system has low voltage DC loads, such as lamps, then it is generally necessary to wire these via the controller. For this purpose it is necessary to select a controller with a battery protection or low voltage disconnect facility which automatically disconnects the loads. The load or output current is determined by calculating the maximum number of appliances which are likely to be switched on at the same time and adding up the current consumption of them all.

5.5.3.4 Battery type

If it is decided to use sealed batteries, then it is important that the controller chosen is suitable. Vented batteries require a higher voltage 'boost charge' periodically which would damage a sealed battery. Suitable controllers have a facility which disables this boost charge.

5.5.4 Inverter

There are four things to consider when choosing an inverter; input, output, power rating and waveform.

5.5.4.1 Input

As with the controller, the input voltage of the inverter is determined by the system voltage. Larger inverters tend to be more expensive or have a lower rating in 12 Volt versions than 24 or 48 Volt, so it may be worth considering a higher system voltage if there is a lot of AC load.

5.5.4.2 Output

The output voltage and frequency are determined by the input voltage of the appliances that the system is designed to power. This is generally determined by the mains supply of the country that it is to be installed in or where the appliances were bought. For example, in Europe the output would need to be 230 Volts at 50 Hertz, whereas in the USA it would be 110 Volts at 60 Hertz. It is very important that this is correct.

5.5.4.3 Power rating

The power rating is the maximum continuous power that can be supplied to the loads. This is determined by adding up the power consumption in Watts of all the appliances that are likely to be switched on at any one time. Many inverters have a large overload capacity, which means that they can provide substantially more than the rated output for short periods of time. This is useful where motors have to be started, particularly in refrigeration systems.

5.5.4.4 Waveform

The choice between a modified sine wave and pure sine wave inverter is not straightforward. The advantages of a modified sine wave are:

- Low cost

- High overload capacity

- High efficiency

And of a pure sine wave are:

- Low noise

- Compatible with all appliances

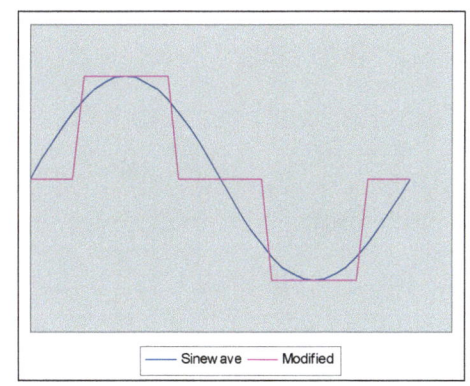

figure 24: inverter waveforms

Recent advances have reduced the cost and increased the efficiency of pure sine wave inverters so that the differences are less clear-cut, especially in the case of larger inverters.

The types of appliances that are incompatible with a modified sine wave inverter are those with a crude power supply, such as certain small battery chargers and those requiring a noise-free supply, such as music equipment. Motors may be noisy.

5.5.5 Appliances

The selection of most appliances will already have been considered during the initial estimation stage of the design process. With the information now available it should be possible to decide upon the actual appliances to be used.

In the case of grid-connected systems, 12 Volt appliances will be unsuitable, and it's likely the most of the appliances will already be in place. However, significant savings can be made by replacing lighting with the most efficient types available, and, when other appliances need replacing, choosing those with the highest possible energy efficiency rating.

5.5.5.1 Lighting

There are many types of lamp to choose from, but the first choice is whether the lighting is to be low voltage DC or mains voltage AC. The advantages of choosing AC are:

- Reduction in cabling cost

- Local availability of lamps

- May use existing wiring

figure 25: types of lamp

Whereas the disadvantages are:

- Cost and complexity of inverter
- Safety in places where electricity is unfamiliar
- Higher power consumption

Generally, AC lighting will normally be chosen in larger systems, where an inverter is already required for other loads.

Since the first issue of this book, LED lighting has become commercially available at reasonable cost. LED lamps are more efficient than even fluorescent strip lights and 'low energy' compact fluorescent lamps, and therefore should be the first choice for any solar power system. If there is no LED lamp which is appropriate then fluorescent lamps are a good option. Any type can be used, including low-energy light bulbs and strip lights, but if strip lights are chosen make sure that they are fitted with an electronic ballast.

There are a variety of low voltage DC lamps available owing to their use for leisure purposes such as boating and caravanning. For reasons of efficiency the first choice should again be LED, followed by fluorescent lamps. These are available in a wide range of styles, including strip-light and 2D types.

Where a spotlight is required an LED is also likely to be suitable. There are fittings which are designed for low voltage application available from caravan and boat suppliers. Another option is to use LED lamps in fittings designed for 'dichroic' halogen lamps. These are normally used with 12 Volts AC from a transformer, but they will work just as well from 12 Volts DC. If they are used in an AC system then an electronic transformer is better than a conventional one.

Despite their low cost, standard incandescent light bulbs should never be used, either in AC or DC applications. Their power consumption is 5 times that of a compact fluorescent lamp for the same light output, so would need five times as many modules and batteries to supply it.

5.5.5.2 Refrigeration

The choice of a refrigerator for solar power operation is subject to a particular set of conditions unlike any other appliance. The power consumption of a refrigerator is quite low when compared with many other devices, but the fact that it has to be powered continuously means that its energy consumption is quite high.

5.5.5.2.1   DC refrigerators

There are three basic types of DC refrigerators; compression, absorption and Peltier effect.

Peltier effect coolers use a solid state heat-pump and are usually small, designed for use in cars. They are quite inefficient, consuming a lot of power for their size, and are really only suitable for mobile applications such as motorhomes where they will be powered by the alternator on the engine most of the time.

The absorption type of fridge is commonly found in caravans, owing to its ability to operate from different energy sources, usually AC and DC power and bottled gas, and also in remote medical centres. While this type of fridge is quite effective when operating from gas or kerosene they should not be considered for constant electrical operation as their performance is poor.

Compressor type fridges are by far the most efficient, especially in low-voltage form. They are more expensive than most other types, but should be the first choice as the extra cost will be repaid many times by their much lower energy requirement.

5.5.5.2.2   AC refrigerators

An AC refrigerator should only be considered if a suitable DC compressor refrigerator is not available, or for grid-connected applications. They are generally less efficient than their DC alternatives especially when inverter efficiency is taken into account.

A further consideration is that the starting current of an AC fridge compressor tends to be very high compared to its DC counterpart, and this can lead to problems operating from an inverter. Advice should be sought from the inverter manufacturer as to whether their inverter is suitable to operate the intended appliance.

5.5.5.3 Microwave ovens

If a microwave oven is to be part of the system, it should be noted that the rated power of the oven is usually the output power and shouldn't be confused with the input power. This is normally to be found on the rating plate on the rear of the appliance and is significantly higher.

Low voltage microwaves are now available and may prove to be more suitable, despite their lower output power which may mean longer cooking times.

5.5.5.4 Other appliances

Other appliances should be chosen primarily on the basis of their power consumption. Greater emphasis should be given to those appliances which are likely to require the most energy, for example those with a high power rating or those which will be on for the longest period. See appendix 5 for a list of common appliances and their power ratings.

5.6 Wiring

5.6.1 Wiring diagram

A wiring diagram should be drawn for even the simplest system, as it helps to ensure that nothing has been overlooked. It will also assist with the cable sizing process and will be essential to the installation process.

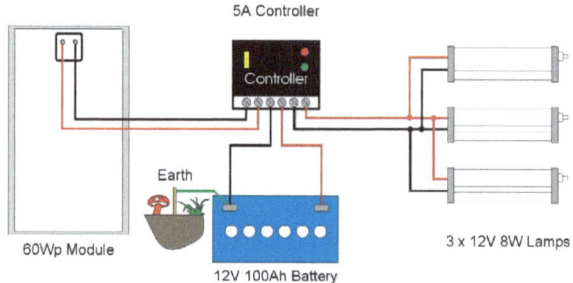

figure 26: example wiring

The wiring diagram will be different for every system and is drawn with reference to the manufacturers instructions for the various system components. If they are not all available at first, draw a general diagram and fill in the details later.

Some example wiring diagrams are included in appendix 4, and the following rules will assist:

5.6.1.1 Array

If the system voltage and the module voltage are the same (usually 12 Volts) then all the modules are wired in parallel. If the system voltage is greater than the module voltage then the modules are wired in strings (pairs in series for 24 Volts, four for 48 Volts) and then the strings connected in parallel.

Array connections for grid-connected systems are made differently. As the output is much higher voltage, they are connected in series up to the maximum input voltage of the inverter. If more modules are required then another series string is connected in parallel with the first, up to the maximum input current of the inverter. This means that there are only certain configurations which are suitable for any particular inverter, and the exact number of modules determined by the sizing process may not be able to be accommodated. A disconnect will be incorporated.

5.6.1.2 Batteries

The batteries are wired in the same way as the array. For example, a number of 12 Volt batteries can be wired in parallel in a 12 Volt system or a number of 2 Volt cells in series to produce 12, 24 or 48 Volts.

www.solar-power-answers.co.uk

5.6.1.3 Controller

A typical controller will have connections to the array, battery and DC loads. It will normally be mounted close to the battery as measurement of the battery voltage is important to its operation.

5.6.1.4 Inverter

If there is an inverter in the system, it is connected directly to the battery using substantial cables.

The synchronous inverter in a grid-connected system is connected to the modules and the incoming mains connection via a disconnect and meter. In many countries, it is a legal requirement that this connection is made by a qualified electrician.

5.6.1.5 Appliances

Lights and other appliances are either connected to the inverter output, if they are mains voltage, or to the controller output if they are DC. This is to take advantage of the low-voltage disconnect voltage function of the controller. There are certain appliances, notably low-voltages refrigerators, which must be connected directly to the battery. Consult the manufacturer's instructions for more information.

5.6.1.6 Circuit protection

Circuit protection is essential in any system to avoid fire caused by the high currents that a battery can deliver into a short-circuit. The basic principles are the same for both AC and DC systems, but in a DC system there is the added requirement to protect the user from electrocution.

5.6.1.6.1 DC systems

In a simple DC system where all the loads are connected via the controller, then the fuse incorporated into the controller may be sufficient. In larger systems it is necessary to incorporate a fuse on the battery positive terminal and ensure that all current from the battery has to pass through it.

In larger DC systems it may also be necessary to incorporate separate fusing for individual circuits. This protection can be normal 230 Volt fuses and circuit breakers for 12 and 24 Volt systems but higher voltages must use special DC fuses.

Provision should be made to earth the battery negative terminal in order to ensure that the system does not float at high voltages.

5.6.1.6.2 AC systems

The requirements for battery fusing are the same as in larger DC systems, but the loads should be fused using a normal 230 Volt consumer unit. This should incorporate a Residual Current Device to protect the users of the system. A good earth should be provided to the consumer unit in addition to that provided at the battery.

5.6.1.7 Wiring accessories

For AC systems, the wiring in the building should follow normal wiring practices, with reference to any regulations in force in the country in which the system is to be installed.

For DC systems where there is any wiring beyond the controller, the same applies with the following differences:

- Larger cable should be used wherever possible, for example 2.5 mm^2 cable for lighting circuits instead of 1.5 mm^2.

- Standard sockets should not be used to connect low-voltage appliances, because of the risk of them being connected to the mains in error. Plugs and sockets designed for DC should be used, or an obsolete system such as the British 15 Amp round-pin plug.

- Standard mains wiring accessories such as light switches and junction boxes can normally be used at voltages up to 30 Volts DC. Consult the manufacturers if in any doubt.

- A separate earth connection to each appliance is normally unnecessary as the negative is earthed.

5.6.2 Cable sizing

Unlike mains wiring, low voltage DC systems lose a significant part of the generated power in the cabling. That's because a lower voltage means that the current is higher, and the power dissipated in the cable is proportional to the square of the current (section 2.1). This means that it is sometimes necessary to use a larger cable than is necessary to carry the current.

If you are using the sizing spreadsheet, enter the lengths of the cables you want to calculate the sizes of into the fields on the *Design Data* page and the results are displayed on the *Components* page. See figure 27.

If you are not using the spreadsheet, then calculate as follows:

$$A = L \times I \times 0.04 / V$$

Where:

A = Cross-sectional area of the cable in mm^2
L = Length of cable in metres
I = Current in Amperes
V = Maximum permissible voltage drop in Volts

The maximum permissible voltage drop should be 5% of the system voltage, that is 0.6 V for a 12 Volt system and 1.2 V for a 24 Volt system.

Cable is available in various sizes depending on the country you are in. The cable used should be the same size or larger then the result of the calculation, never smaller. See appendix 3 for a conversion table between metric and other systems of cable sizing.

Cable lengths		
Array	10.0	m
Inverter	1.0	m

Lamp	10.0	m
Computer		m
Television		m
Radio		m
Fridge	5.0	m
Other		m
		m
		m
		m
		m
		m
		m
		m

Cabling		
Array	16.3	mm^2
Inverter	0.0	mm^2

Lamp	0.6	mm^2
Computer	0.0	mm^2
Television	0.0	mm^2
Radio	0.0	mm^2
Fridge	1.4	mm^2
Other	0.0	mm^2
	0.0	mm^2
	0.0	mm^2
	0.0	mm^2
	0.0	mm^2
	0.0	mm^2
	0.0	mm^2
	0.0	mm^2

figure 27: cable sizing

6 Installation and Commissioning

Before commencing the installation it is important to familiarise yourself with the manufacturer's instructions supplied with each of the components. The site visit will have allowed you to identify a mounting position for each item. It will be helpful to draw a wiring diagram before starting the installation. Every system will be different, but see appendix 4 for example wiring diagrams.

There is no correct order for the installation of the various system components, only their eventual connection and commissioning.

6.1 Safety

At all times during the installation, the safety of the installers and public must be paramount. Keep the public away from the installation site at all times, using barriers or fencing where necessary. Pay particular attention to the safety of children.

6.1.1 Electrical

Although solar power systems are generally low voltage, always observe the wiring regulations for the country of installation. Bear in mind the following:

- Inverter output is mains voltage AC and can be lethal. Treat as for any other mains supply.

- Solar arrays generate electricity when exposed to the sun, whether connected to control equipment or not. Treat array output cables as live and cover array when making connections.

- The open circuit voltage of a solar array is significantly greater than the system voltage. For example a 48 Volt array can have an open circuit voltage of nearly 90 Volts, which can be lethal to children, the elderly or anyone with a heart condition.

- Batteries can produce currents of hundreds or even thousands of amps giving rise to the risk of fire. Take great care to protect the battery terminals from shorting by tools and remove all jewellery.

- The DC input voltage of grid-connected synchronous inverters can be hundreds of volts, and higher than the mains voltage.

If in any doubt about your abilities, or if required by local regulations, then a qualified electrician must be employed.

6.1.2 Chemical

Lead acid batteries contain dilute sulphuric acid and liberate hydrogen when charging. Observe the following precautions:

- Take great care when filling batteries with electrolyte; wear suitable protective clothing including eye protection and carry out in a well ventilated area, preferably outdoors.

- Do not smoke near batteries and ensure room is well ventilated.

- Take care to prevent arcing near battery terminals as explosion may result.

- Keep first aid and eyewash equipment close at hand when working on batteries.

6.1.3 Handling

Batteries and solar arrays present certain hazards in handling as follows:

- Lead acid batteries are extremely heavy. Use appropriate lifting gear and ensure adequate help is available.

- Most solar panels are made from glass. Treat as fragile.

- Installing arrays may involve working at height. Observe all necessary precautions and employ the services of a qualified rigger or roofer if necessary.

6.2 Array

The installation of the array is determined by the type of support structure, but is normally completed in three stages. The photographs accompanying this section were taken during a solar power course in Angola, and illustrate some of the difficulties faced when certain tools and equipment are unavailable.

6.2.1 Assembly

Before fixing to the roof or ground the modules must be mounted to the support rails to create a single unit. In larger systems the array may be split into sub-arrays. Each of these is treated as a separate array.

If a proprietary support structure is used then the correct nuts and bolts will be supplied. Otherwise use high-tensile nuts and bolts the correct size for the mounting holes in the modules, and ensure that either locking washers or self-locking nuts are used to prevent loosening as a result of vibration due to the wind.

Lay the modules face down in the correct alignment. Ensure that the ground is flat; a grassed area is ideal. Place a blanket or similar on the ground to avoid any damage to the glass. Bolt the mounting rails to the modules, taking care to ensure that no strain is placed upon the modules. Do not put any weight on the modules themselves.

6.2.2 Connection

It is usually easiest to connect the individual modules together and connect the long output cable to them before putting them into position, especially if they are to be roof mounted. Connect all the modules together first; either in one, two or four groups depending on the system voltage and then connect the output cable. Check that all connections are sound before replacing the terminal box covers as this is hopefully the last time you will see them.

6.2.3 Testing

It is best to test the array before mounting to avoid having to take it down again. To do this, turn it over and lay it on the ground face up. Make sure you have enough people helping to avoid twisting or dropping the array.

Using a multimeter set to an appropriate DC Voltage range, measure the voltage between the two cores of the output cable. It should be equivalent to the open-circuit voltage of one module multiplied by the number of modules in series. That is, about 20 V for a 12 V system, 40 V for a 24 V system and 80 V for a 48 V system. 80 Volts is a dangerously high voltage, and 40 Volts is enough to give a nasty shock in the wrong circumstances; take care when performing these measurements and cover the array with an opaque material before connecting the meter if in any doubt.

figure 28: array testing

Exercise extreme care if testing an array for a grid-connected system, as the modules are connected in series and therefore the voltage is much higher.

Once this test is complete it is advisable to make the array safe before lifting it into place. There are two ways of doing this; either cover it with an opaque material or connect the cores of the output cable together to short-circuit the array. This will not cause any damage and is my preference. As grid-connected arrays are normally connected with insulated connectors, this step will not be necessary.

figure 29: lifting array

www.solar-power-answers.co.uk

6.2.4 Mounting

Once you are certain that the array is correctly assembled it is time to lift it into place and secure it. Usually there are a number of rails to be mounted to the roof or foundations first. Ensure that these are orientated correctly so that the array will point towards the equator once mounted; that is towards the south in the northern hemisphere and towards the north in the southern hemisphere. Use appropriate fixings and follow safe working practices for working on roofs where appropriate.

Assemble a team of sufficient number to lift the array into place without bending it. Often the best method is to place two ladders parallel to each other and walk up them with the array in between. Often there is only one ladder in the village though, so you may have to improvise. See figure 29 – this was an installation in an Angolan village where, despite ladders being promised, none were available. Abandoning the installation wasn't appropriate but fortunately there were enough people available to make it possible to lift the array directly onto the low roof.

Carefully lift the array into position on the mounting rails. Assemble using the correct nuts and bolts. If the mounting structure is adjustable for tilt, now is the time to set it. Support the array whilst making any adjustments and set the tilt based on the latitude of the location as follows:

- Optimised for winter, e.g. lighting systems in temperate zones: Angle of latitude +15° from the horizontal.

- Optimised for summer, e.g. holiday homes: Angle of latitude -15° from the horizontal.

figure 30: mounted array

- All year round performance, e.g. medical systems in the tropics: Angle from the horizontal equal to angle of latitude.

Do not set the tilt to less than 10° from the horizontal whatever the latitude, as a tilt of less than this will allow dirt to build up on the array which will reduce its performance.

6.3 Battery

Installation of the battery may be as simple as taking a wet-charged or sealed battery out of a box and placing it on a firm and level surface. Alternatively it may involve mixing acid to the right concentration and filling the batteries on site.

6.3.1 Siting

The batteries need to be mounted such that they are secure, i.e. they can't fall over, they are protected from unauthorised access and away from sources of ignition. The room or container that they are in should be ventilated so as to allow the hydrogen produced by charging to escape. This applies even to sealed batteries as they are able to vent excess gasses should the charging system malfunction.

Practically this is most likely to mean one of two things:

1. On a solid floor or racking within a locked and well ventilated room.

2. In a purpose designed battery box.

It is important that it is possible to gain access to the batteries in order to perform maintenance. In the case of a sealed battery this means the terminals, but for a vented battery it may mean access to the level markings on the side and the filling caps.

6.3.2 Commissioning

It may be necessary to commission the batteries either before taking them to site or once at site. Always follow the manufacturer's instructions and, if it is not possible to do exactly as instructed, ask the manufacturer or supplier for advice. Correct battery commissioning is vital to the proper performance of the battery. General instructions are as follows.

figure 31: battery installation

Picture courtesy of Bright Light Solar Ltd

6.3.2.1 Sealed batteries

Sealed batteries of both gel and AGM types are always supplied filled and usually charged. It may be necessary to give them a refresher charge before putting them into service. This should be performed with a regulated mains charger if possible. If this can't be done for any reason then the system should be used as little as possible for the first 24 hours in order to allow the batteries to become fully charged.

6.3.2.2 Wet batteries

When batteries are supplied already filled with electrolyte, they are usually charged and are treated in the same way as described for sealed batteries.

6.3.2.3 Dry-charged batteries

Some wet batteries are supplied dry-charged. This means that they have been charged, the electrolyte emptied out and the battery dried and sealed. It is important that it remains sealed if it is to be stored before being put into service. Once it is required, the seals should be removed and dilute sulphuric acid of the correct specific gravity used to fill the cells to the filling mark. The acid will usually be supplied with the battery. The battery should then be allowed to stand for a period of time to allow any air to escape.

A dry charged battery, once filled, will achieve about an 80% state of charge. For this reason it must be charged before putting into service. Normally this will be specified by the manufacturer as being at a specific voltage for a particular amount of time. Again it should be accomplished by means of a mains operated charger if possible, although it may be possible to use the solar system to provide this charge if the loads are disabled for a certain amount of time.

6.3.2.4 Dry uncharged batteries

Sometimes batteries, normally large cells, are supplied in a dry uncharged state. It is critical that the manufacturer's instructions are followed to the letter as the initial charge is important to the formation of the plate structure. Normally they will be filled as with dry-charged batteries and then subjected to an extended charge, often taking some days. This must be performed with the correct type of charger and it is advisable to avoid purchasing batteries in this state if possible. If wet batteries have to be transported by sea or air the best option is to obtain them in a dry-charged state with the acid in suitable packaging as advised by your shipping company.

6.3.2.5 Mixing acid

It may sometimes be necessary to mix your own electrolyte from concentrated sulphuric acid and distilled water. This should be avoided unless absolutely necessary, for instance in developing countries where it may be impossible to purchase ready made electrolyte and shipping it from the manufacturer is impractical. It is essential that the sulphuric acid and distilled water are of the highest purity.

Before you start collect together the following items:

- Safety clothing, including a chemical splashes apron, face shield and suitable gloves.

- A suitable non-metallic mixing vessel and stirrer. Glass is ideal although you may have to use whatever is available such as a plastic bucket and wooden stick. I write from experience on this.

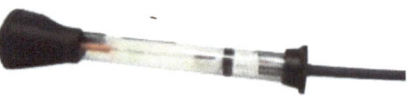

figure 32: Battery hydrometer

- A glass thermometer calibrated from at least 10 - 80° C.

- A battery hydrometer, which can be bought from a battery specialist or tool shop.

- A container to hold the finished electrolyte. A plastic drum is ideal and should be marked "Sulphuric acid. Highly corrosive."

It is important that you always add acid to water and never the other way round. In this way, you are never diluting concentrated acid, which can cause it to boil explosively with very serious consequences. Mix a little at a time, add the acid in small amounts and stir thoroughly. Bear in mind that it will become hot.

Each time you add some acid, check the specific gravity by drawing a little electrolyte into the hydrometer and expelling it, then drawing in enough just to lift the float. Read off the specific gravity from the scale and record. Then measure the temperature of the electrolyte and add 0.001 to the reading on the hydrometer for each degree Celsius above the specified temperature.

For example:

The battery manufacturer calls for an electrolyte with a specific gravity of 1.240 at 25° C. The hydrometer reads 1.230 and the thermometer 35° C. The electrolyte is therefore correctly mixed; 1.230 plus 10 (35-25) times 0.001 is 1.240.

If at any time the electrolyte reaches a temperature of 60° C, go and have a cup of tea until it cools down again. You really don't want it to boil.

Once each batch of electrolyte is mixed, add it to the waiting container until sufficient has been mixed. Allow to cool to room temperature before using. Ensure that it is all mixed thoroughly so that there is no variation between the strength of electrolyte across the cells of the battery. I have achieved this in the past by an eight-hour canoe journey; you may choose to invert the (sealed) container a few times or stir. Be careful if stirring to avoid splashes. Store the container securely if it is not to be used immediately.

6.3.3 Connection

Once the battery has been put in place, the individual batteries or cells can be connected together to form a single battery. If possible use battery interconnect cables supplied by the battery manufacturer or supplier; if you have to make your own then use the thickest cable that is practical.

Take great care when connecting the batteries; remember that they are never switched off and the current from a single cell can be sufficient to heat tools to red heat in moments. Remove all metal jewellery before starting and use insulated tools where available. Do not connect the output cables at this point.

6.3.4 Earthing

Provision should be made to earth the battery negative terminal. If no suitable earth is available, then an earth rod must be driven into the ground outside as near as possible to the battery. This is connected to the battery negative terminal via green and yellow earth cable of at least 2.5 mm^2 and preferably 6 mm^2 cross-sectional area.

6.4 Control equipment

The controller, inverter and any other control and monitoring equipment can now be installed.

6.4.1 Mounting

The control equipment is generally mounted on a vertical surface to aid cooling. See the instructions for your particular units. In a building, the normal position is on the wall above the batteries. Bear in mind that inverters in particular can be very heavy, and ensure that the wall is strong enough to support the weight.

figure 33: control equipment

Picture courtesy of Bright Light Solar Ltd

6.4.2 Wiring

The wiring may now be put in place, following the diagram and cable sizes arrived at in section 5.6. Take great care to observe the correct polarity, and ensure that all connections are well tightened. Use crimp eyelets to connect to stud terminals, and attach them with the correct crimp tool, not a pair of pliers. Clip the cables securely to the wall where possible to keep them out of harm's way and make them look neat. Do not make any of the final positive connections to battery, array or load at this stage.

All wiring must be installed in conformity with local electrical regulations, by a qualified electrician where necessary. See section 5.6.1.7 for recommendations for DC wiring.

6.5 *System Commissioning*

System commissioning is the process of checking and testing the installation and putting it into service. It may be tempting to hurry this procedure; time may be running short and the user may be impatient to see the system working. However, the future reliability of the entire system depends on careful commissioning. If the equipment you are using has any specific commissioning instructions then follow those in preference to the instructions below.

6.5.1 Visual check

With the wiring diagram in your hand, carefully examine the system to ensure that everything is as it should be. Pay particular attention to the polarity of connections and don't forget the battery earth.

6.5.2 Connections

Check the security of all the connections to the control gear and any other connections that have been made already, such as the battery negative and earth connections. Even if you've just connected them, check them again.

6.5.3 Applying power

With any load isolators or circuit breakers switched off, connect the loads to the controller and / or inverter.

Next, measure and record the battery terminal voltage and connect the battery positive terminal. Insert the battery fuse if fitted. Be aware that a spark is likely to occur, so ensure that the room is well ventilated and blow across the battery caps first to clear any hydrogen if the battery is of the vented type. Measure the voltage at the battery connections of the controller and inverter. This should be the same as the battery terminal voltage. If not, check all the connections and the battery fuse.

Finally, connect the photovoltaic array. Before you do this you may want to cover it with an opaque material if possible; remove this as soon as the connection is secure. Measure the voltage at the input terminals of the controller – this should be the same as or slightly higher than the battery terminal voltage.

If there is a reasonable amount of daylight then the controller should show that the battery is charging. If it's gone dark already then you really need to come back tomorrow. Check that the battery is actually charging by measuring the terminal voltage, which should be higher than that initially recorded and rising.

www.solar-power-answers.co.uk

Now switch on the loads. Go round and test them all, if there are more than one, to make sure that they all work. Make sure you switch off anything that is not in use afterwards.

The commissioning process is now complete.

6.5.4 Handover

If you are not to be the eventual user yourself, then an explanation of the operation of the system should be given to the end user before leaving the site.

This explanation should cover the following:

- Principles of operation
- Effect of weather and season on available energy
- Importance of keeping energy usage to a minimum
- Operation of low voltage disconnect
- Meaning of indications on control equipment
- Safety aspects of batteries

The user should also be given copies of the instruction manuals for the components.

7 Maintenance

The requirement for scheduled maintenance is limited to the following:

As Required	Clean Photovoltaic array
6 monthly (or as specified by manufacturer)	Check electrolyte level in vented batteries and top up with distilled water if necessary
12 monthly	Clean battery terminals and protect with petroleum jelly Clean top of battery Check security of all connections

8 Appendices

8.1 Appendix 1 – Insolation Maps

Each of these maps shows the daily insolation on a titled surface for the whole globe for one month.

To use the maps, find your location and use the colour key at the bottom of the map to determine the insolation for that month. To find the design insolation, use the figure from the month with the lowest insolation. If there are any months during which the system will not be used then disregard that data.

For more information see section 5.4.2.1.

Acknowledgement:

These data were obtained from the NASA Langley Research Center Atmospheric Science Data Center Surface meteorological and Solar Energy (SSE) web portal supported by the NASA LaRC POWER Project.

https://eosweb.larc.nasa.gov/sse/

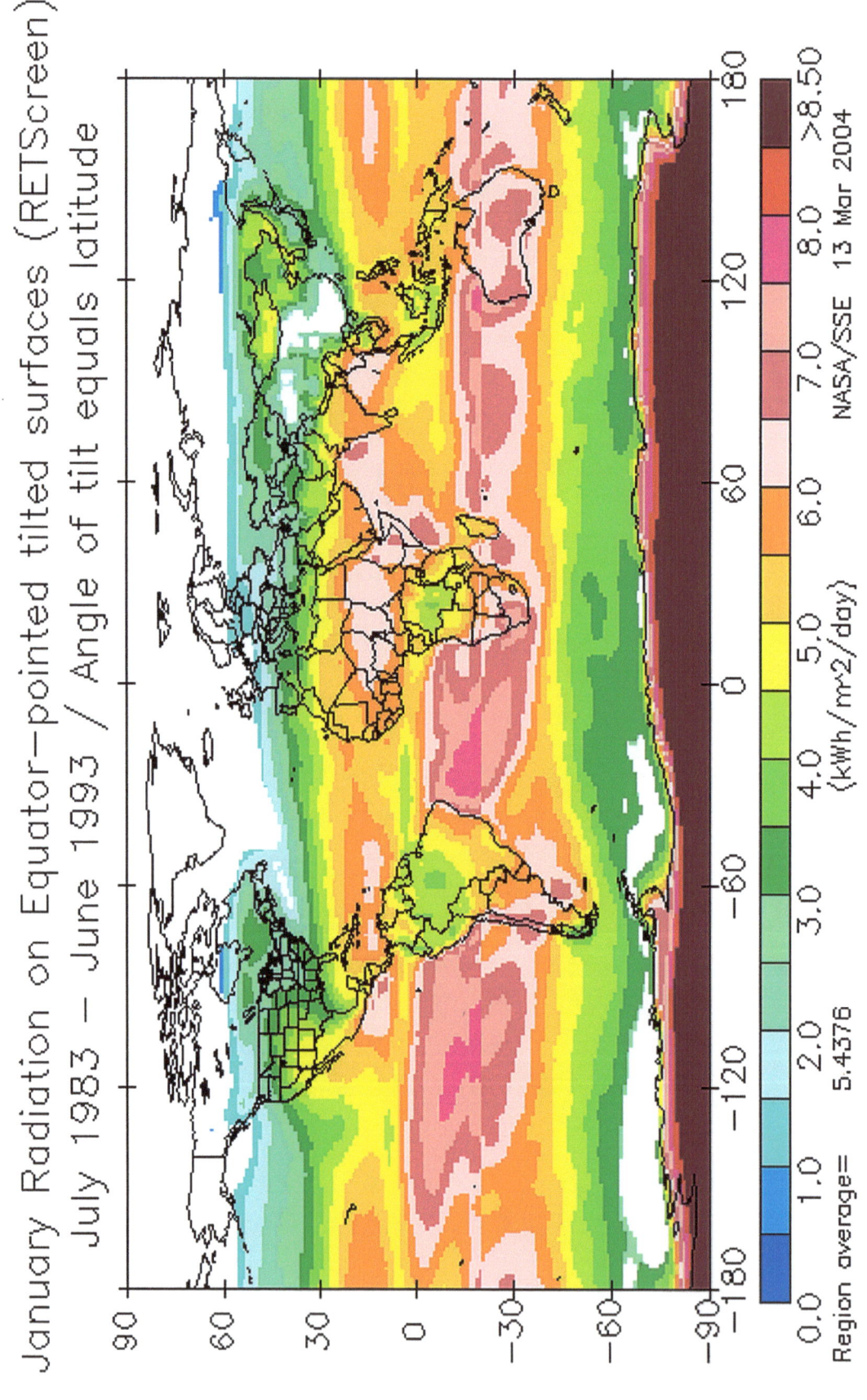

January Radiation on Equator-pointed tilted surfaces (RETScreen)
July 1983 – June 1993 / Angle of tilt equals latitude

NASA/SSE   13 Mar 2004

(kWh/m^2/day)

Region average= 5.4376

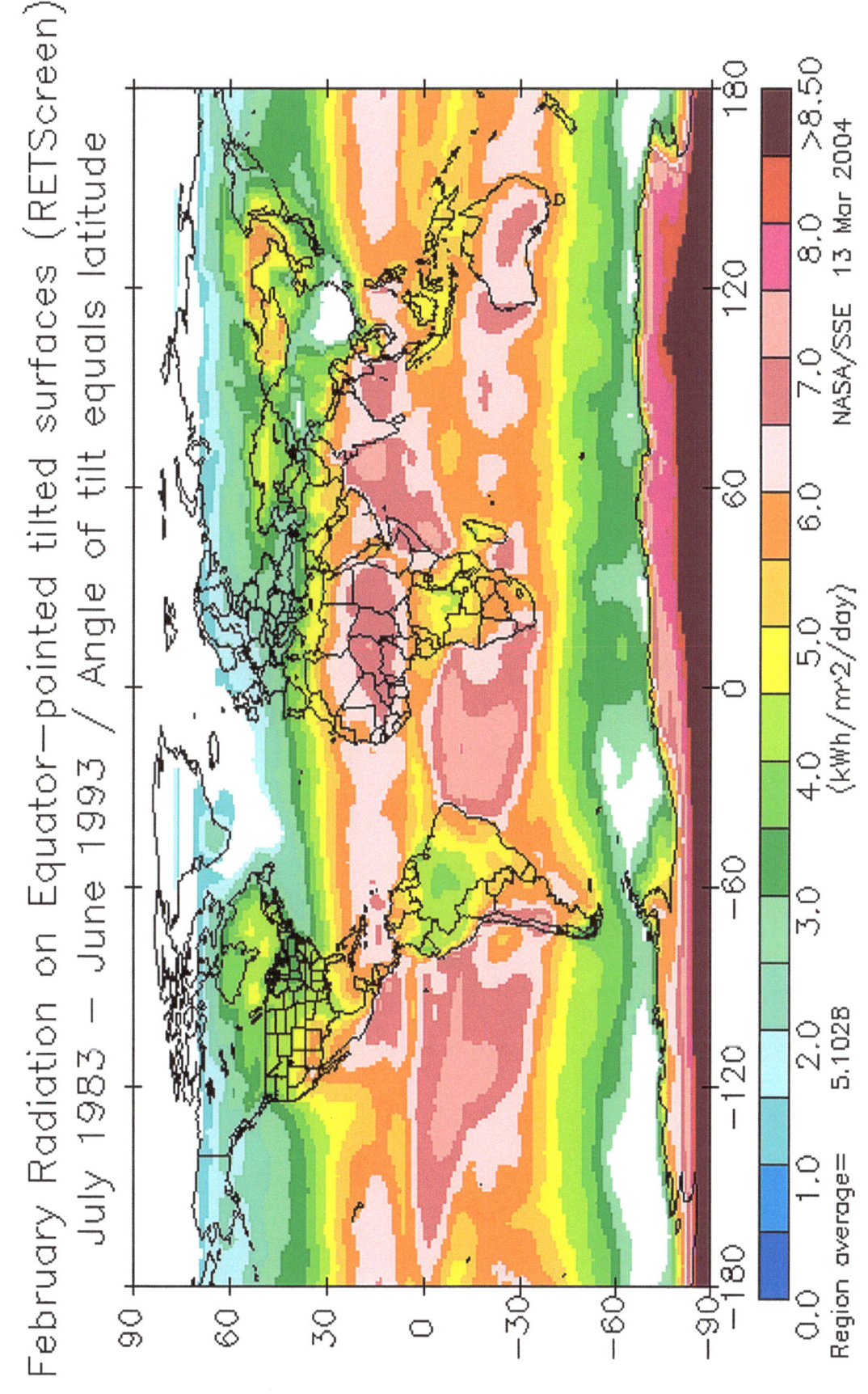

February Radiation on Equator-pointed tilted surfaces (RETScreen)
July 1983 – June 1993 / Angle of tilt equals latitude

www.solar-power-answers.co.uk

March Radiation on Equator-pointed tilted surfaces (RETScreen)
July 1983 – June 1993 / Angle of tilt equals latitude

www.solar-power-answers.co.uk

April Radiation on Equator-pointed tilted surfaces (RETScreen)
July 1983 – June 1993 / Angle of tilt equals latitude

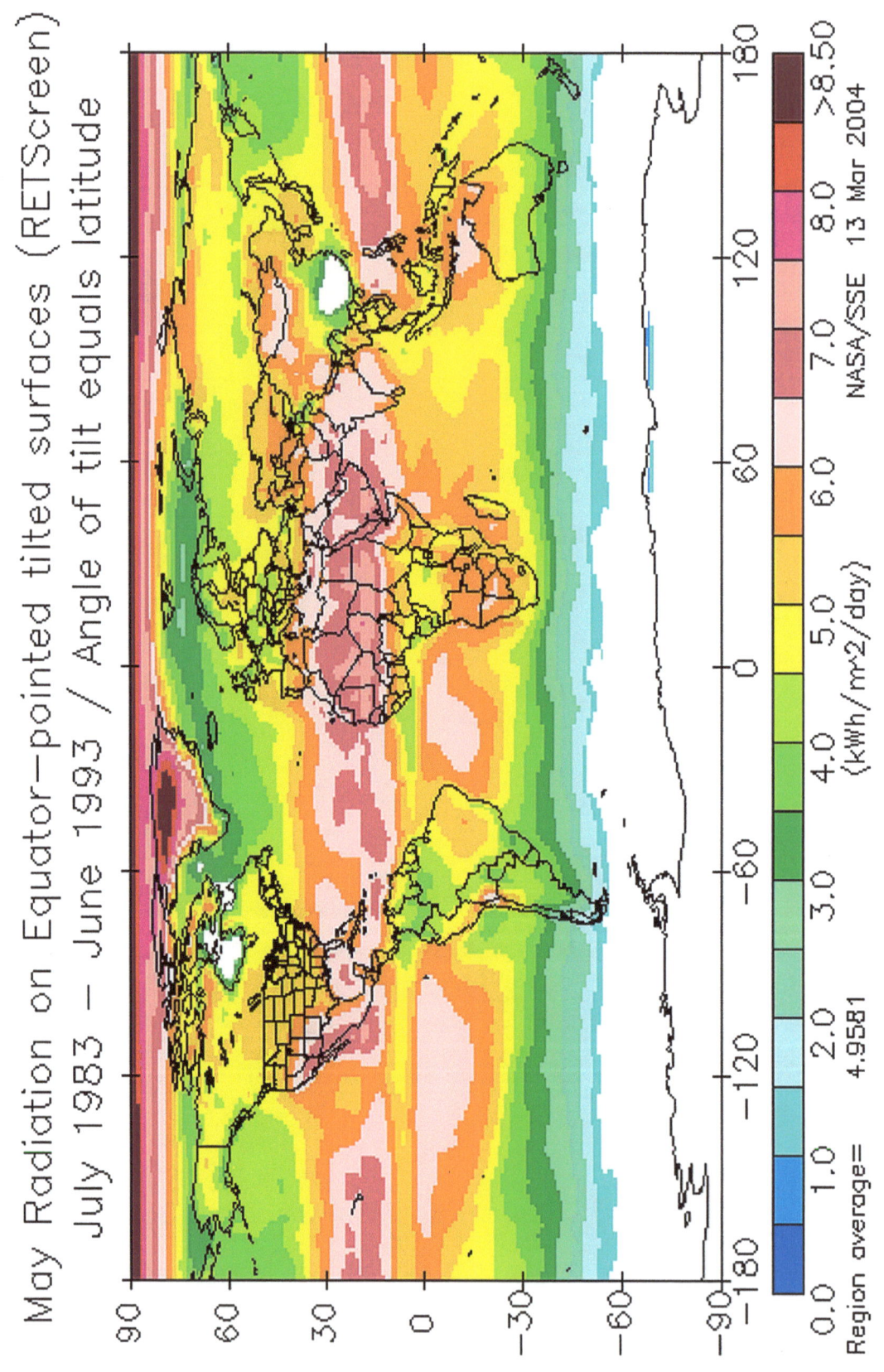

May Radiation on Equator-pointed tilted surfaces (RETScreen) July 1983 – June 1993 / Angle of tilt equals latitude

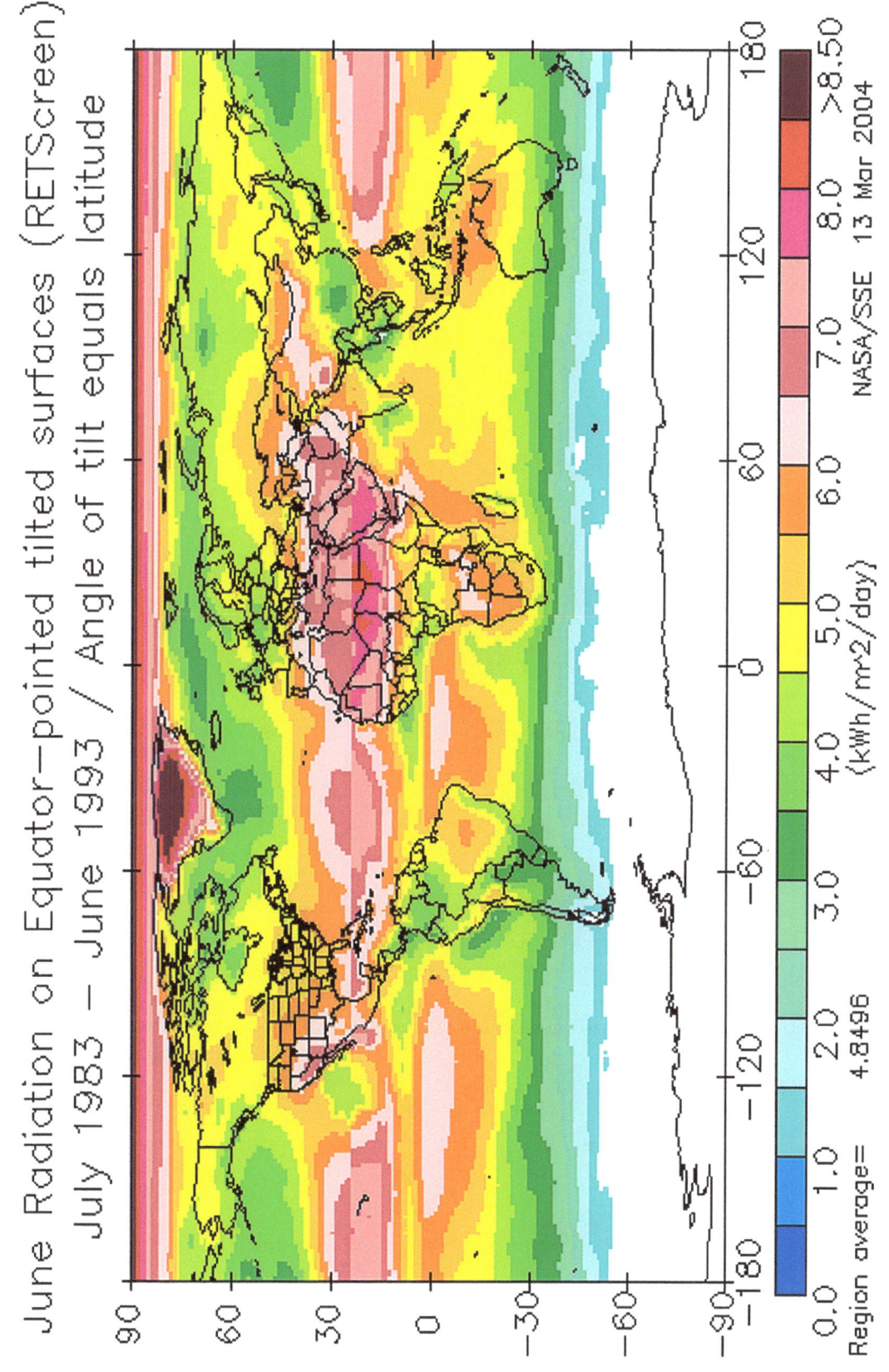

June Radiation on Equator-pointed tilted surfaces (RETScreen)
July 1983 – June 1993 / Angle of tilt equals latitude

NASA/SSE   13 Mar 2004

(kWh/m^2/day)

Region average=   4.8496

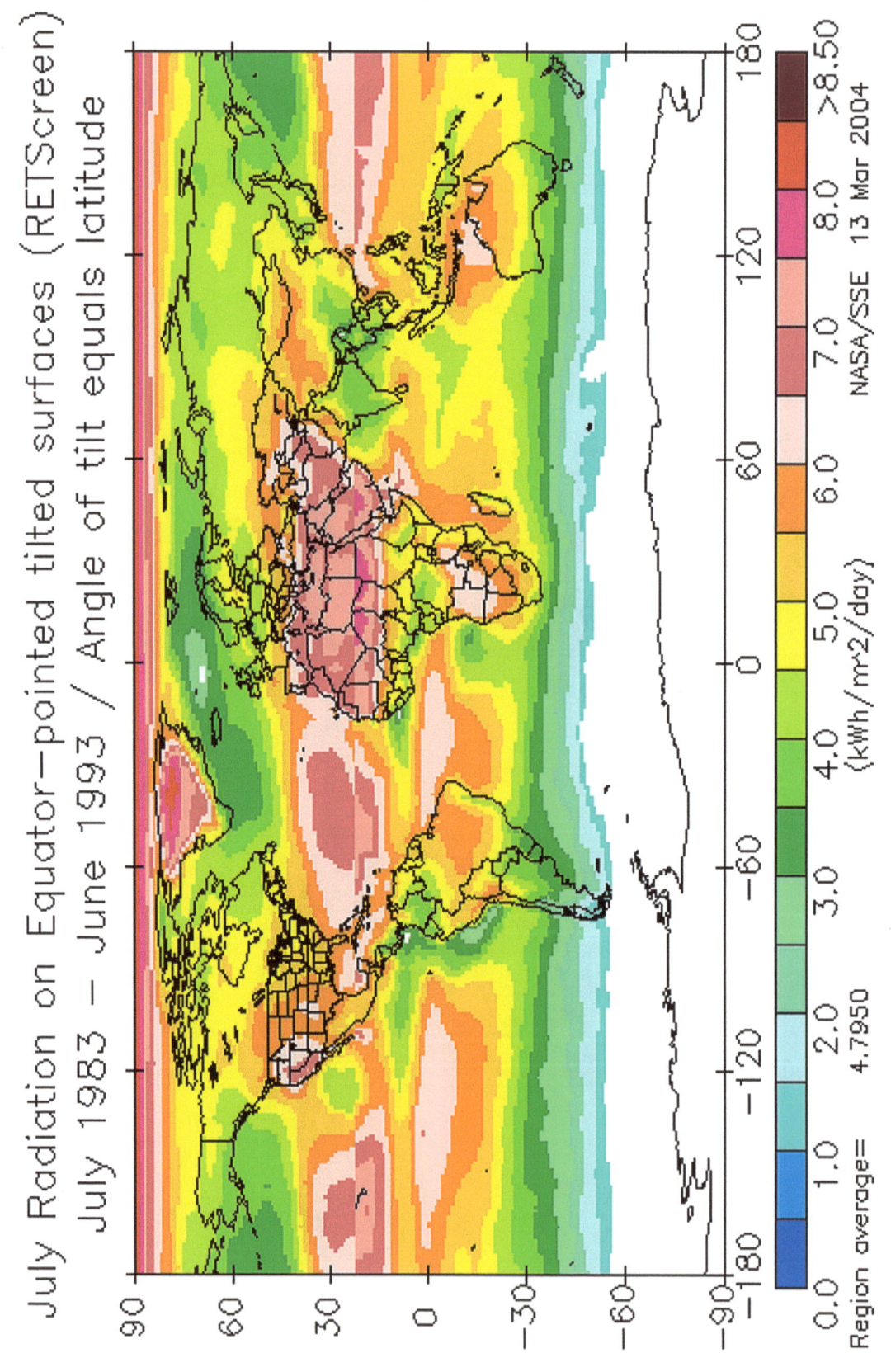

July Radiation on Equator-pointed tilted surfaces (RETScreen)
July 1983 – June 1993 / Angle of tilt equals latitude

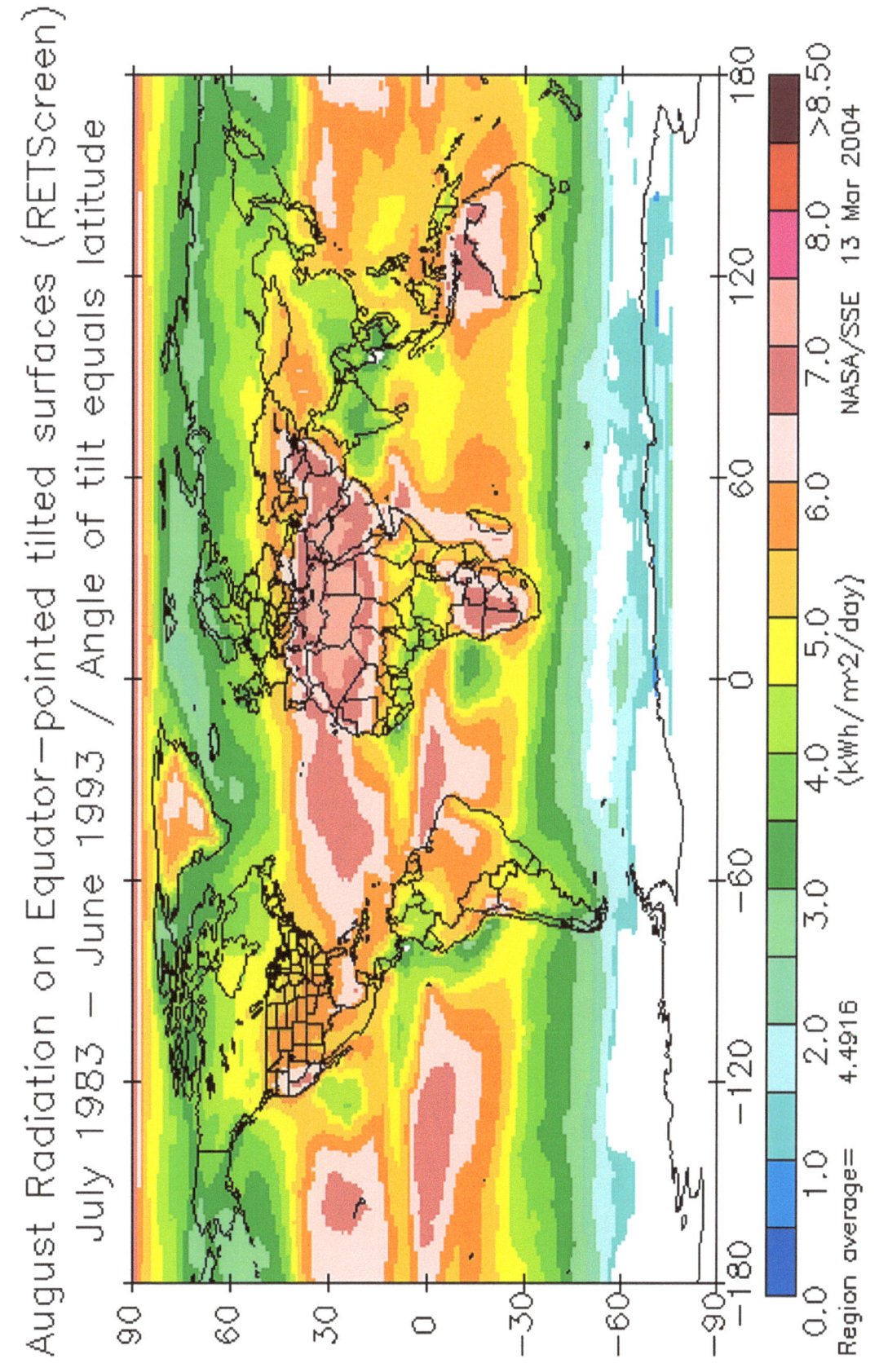

August Radiation on Equator-pointed tilted surfaces (RETScreen)
July 1983 – June 1993 / Angle of tilt equals latitude

www.solar-power-answers.co.uk

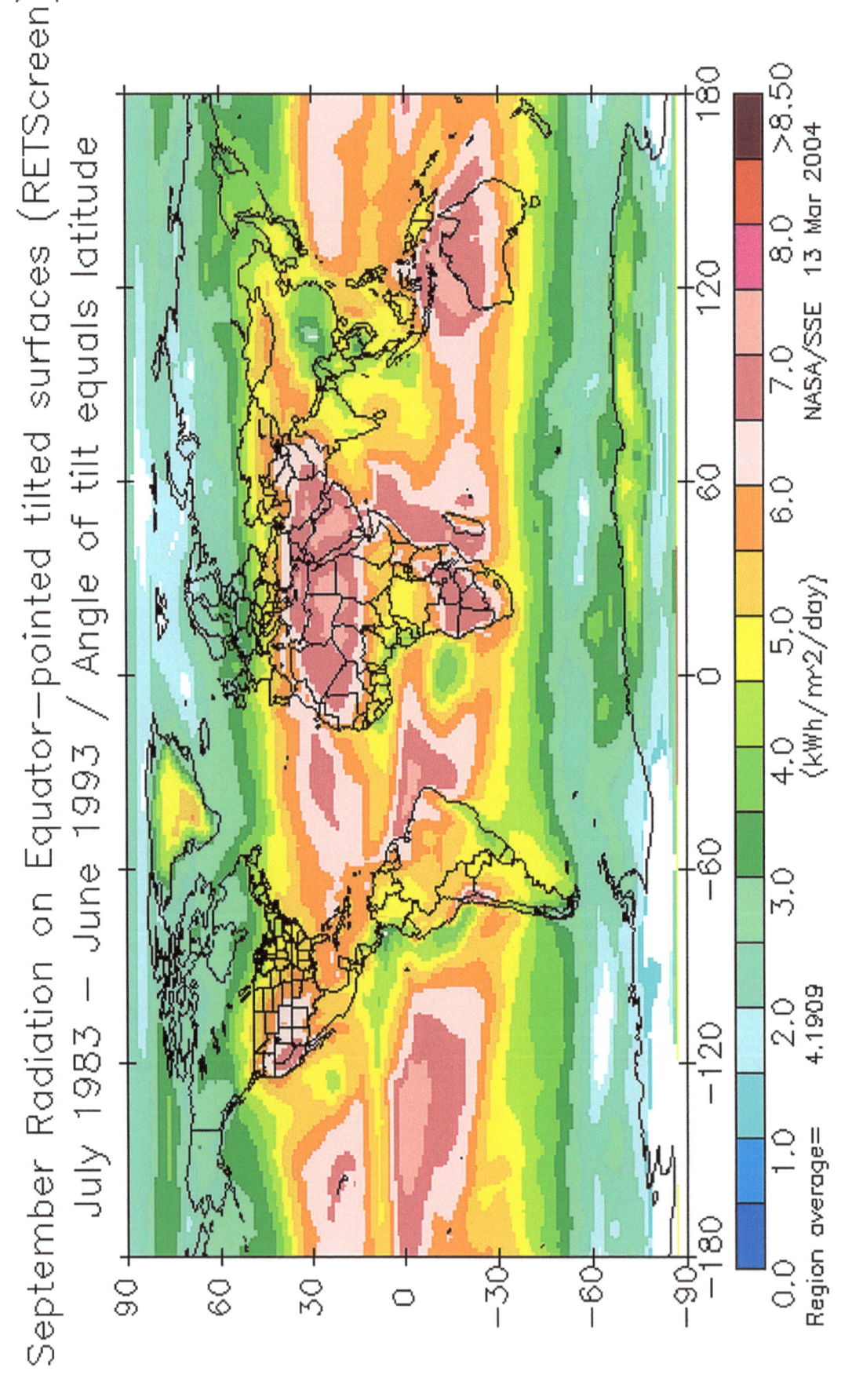

September Radiation on Equator—pointed tilted surfaces (RETScreen)
July 1983 – June 1993 / Angle of tilt equals latitude

NASA/SSE   13 Mar 2004

(kWh/m^2/day)

Region average= 4.1909

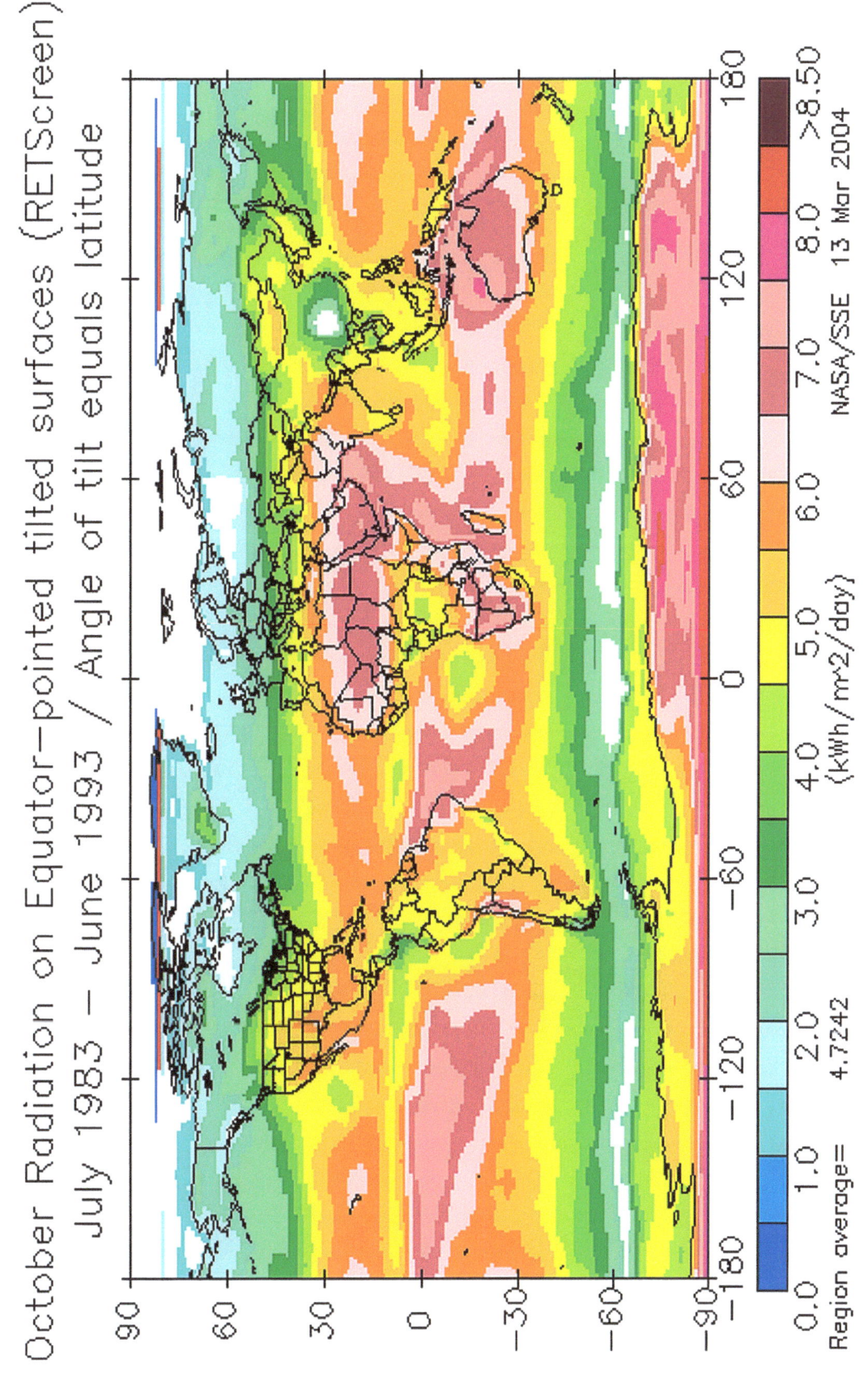

October Radiation on Equator-pointed tilted surfaces (RETScreen)
July 1983 – June 1993 / Angle of tilt equals latitude

63

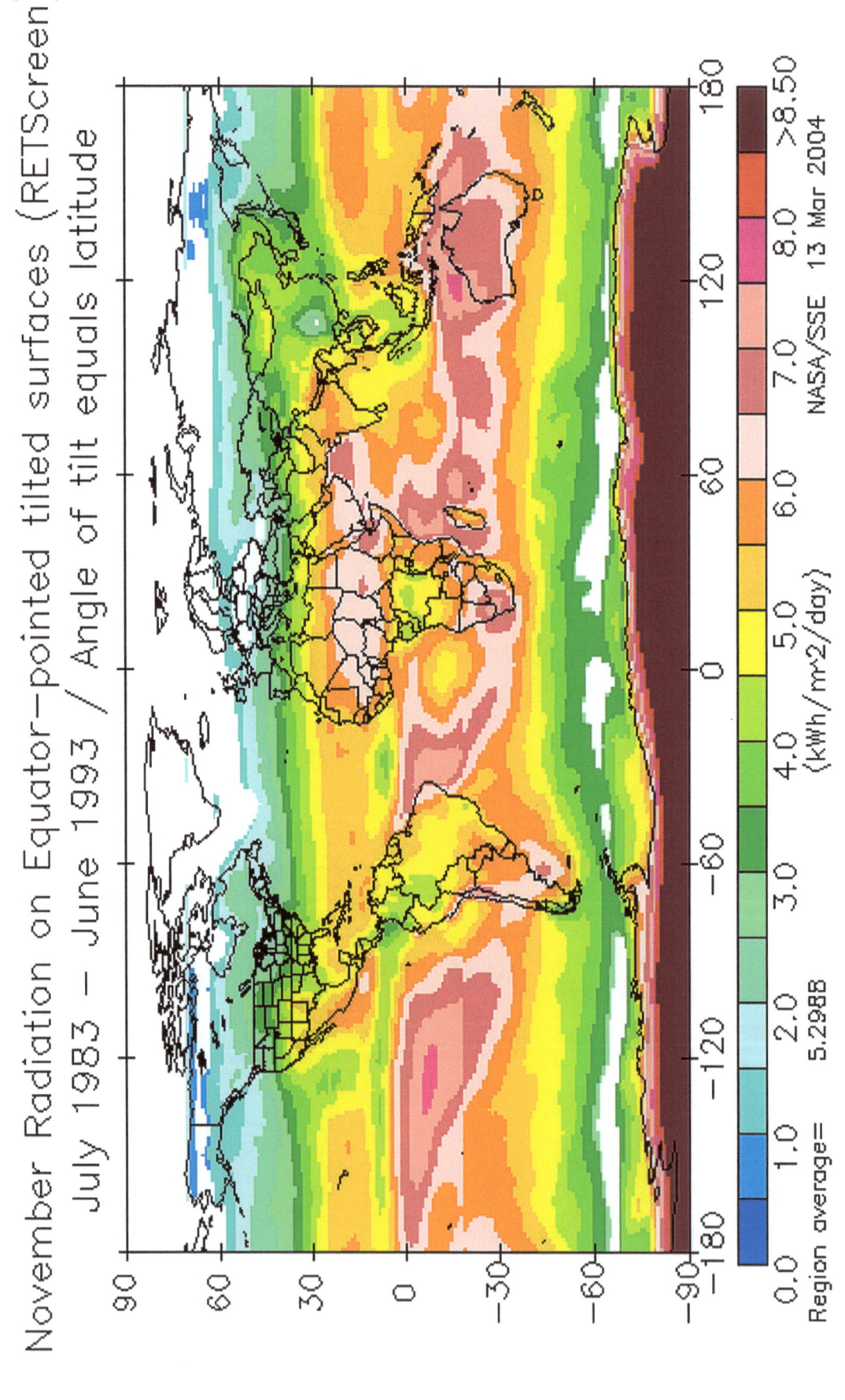

November Radiation on Equator-pointed tilted surfaces (RETScreen)
July 1983 – June 1993 / Angle of tilt equals latitude

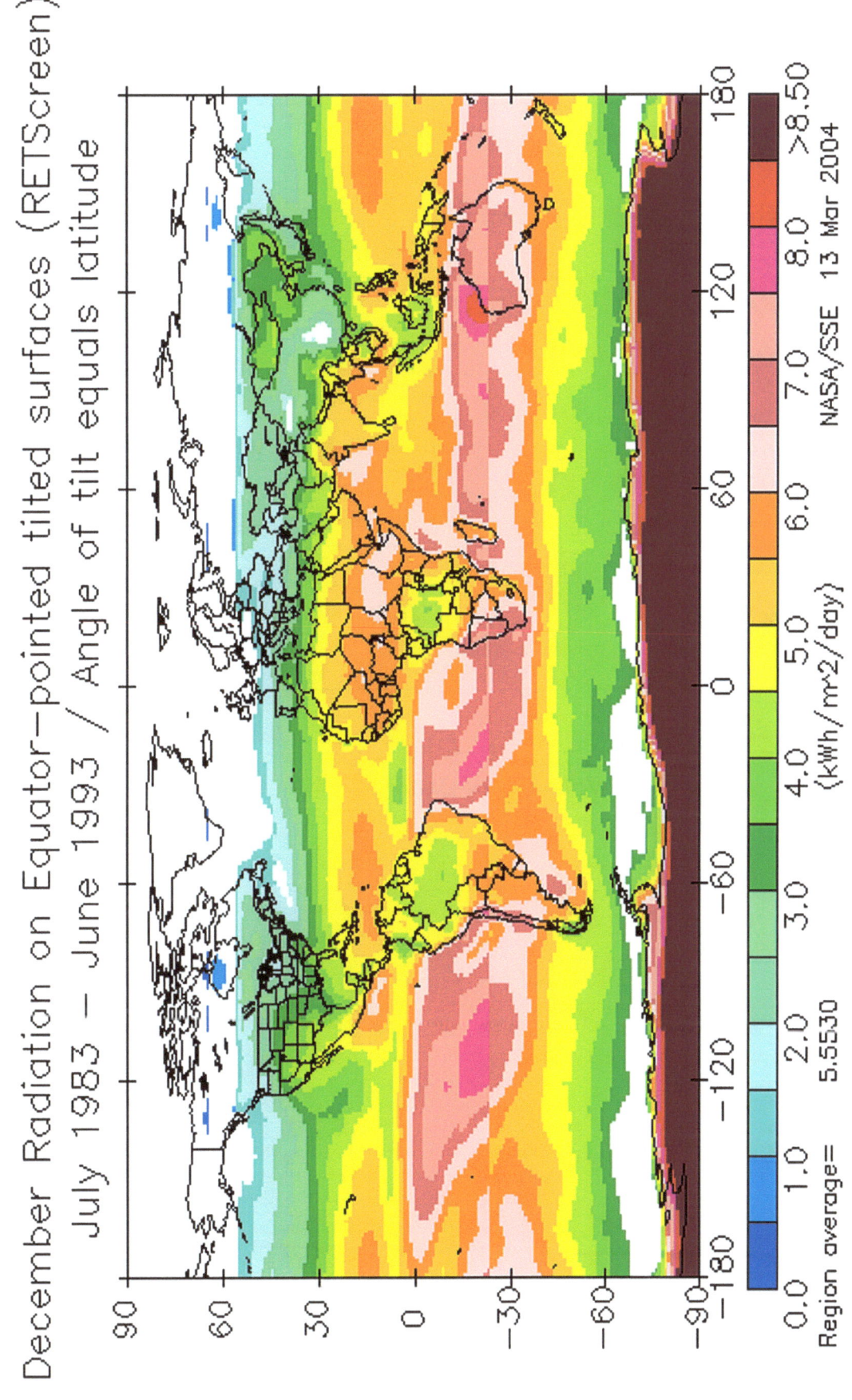

December Radiation on Equator-pointed tilted surfaces (RETScreen)
July 1983 – June 1993 / Angle of tilt equals latitude

65

8.2 Appendix 2 – Battery Voltages

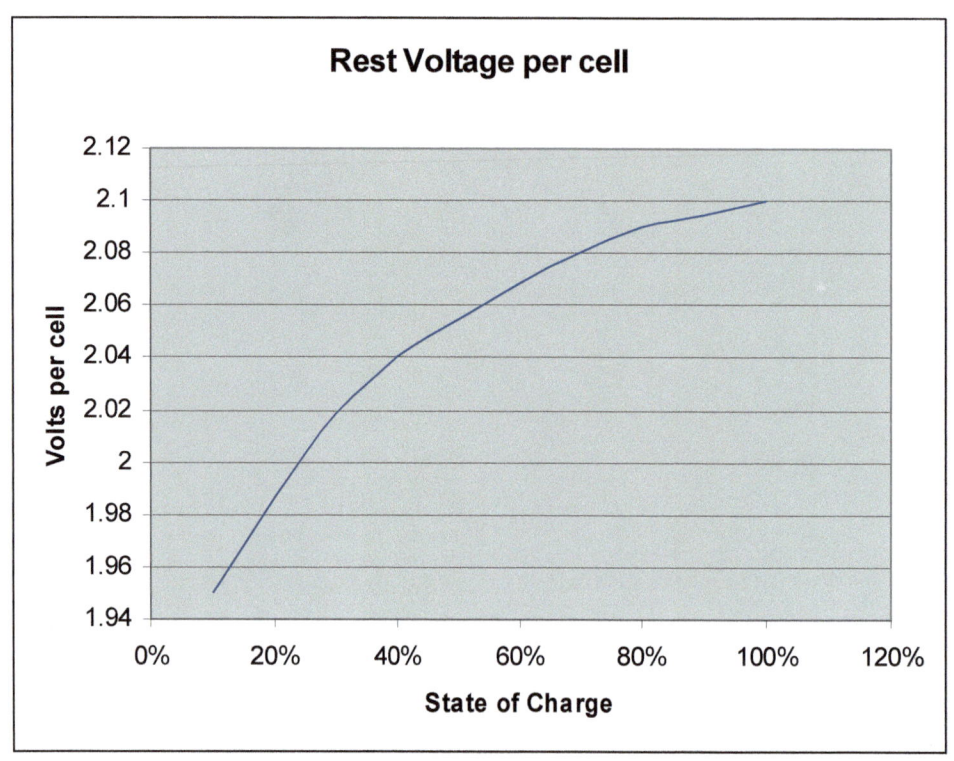

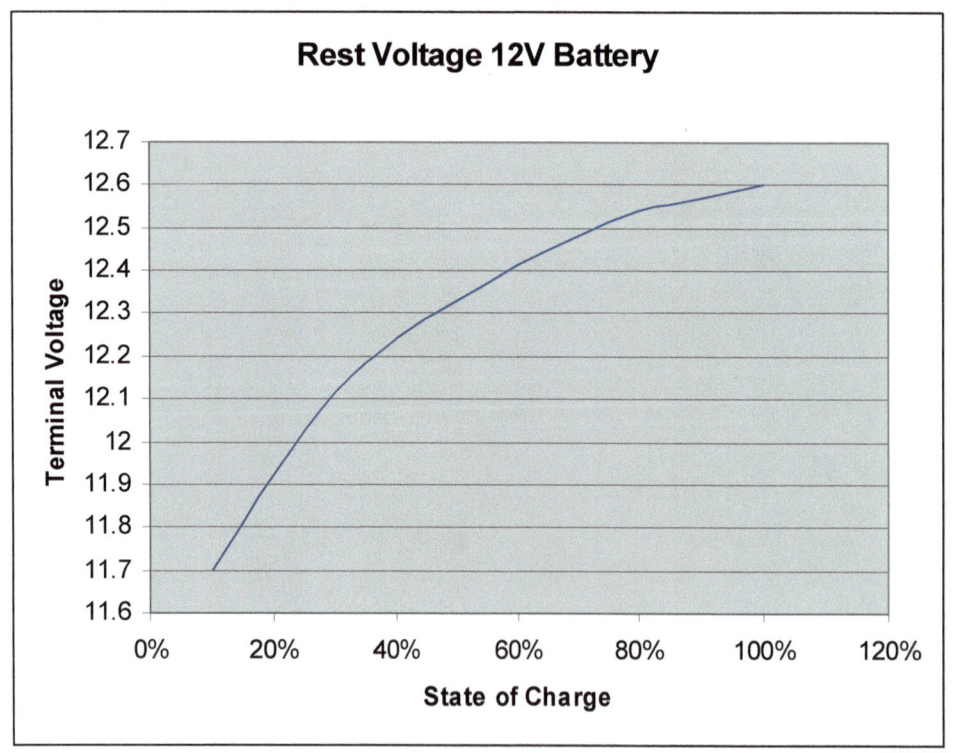

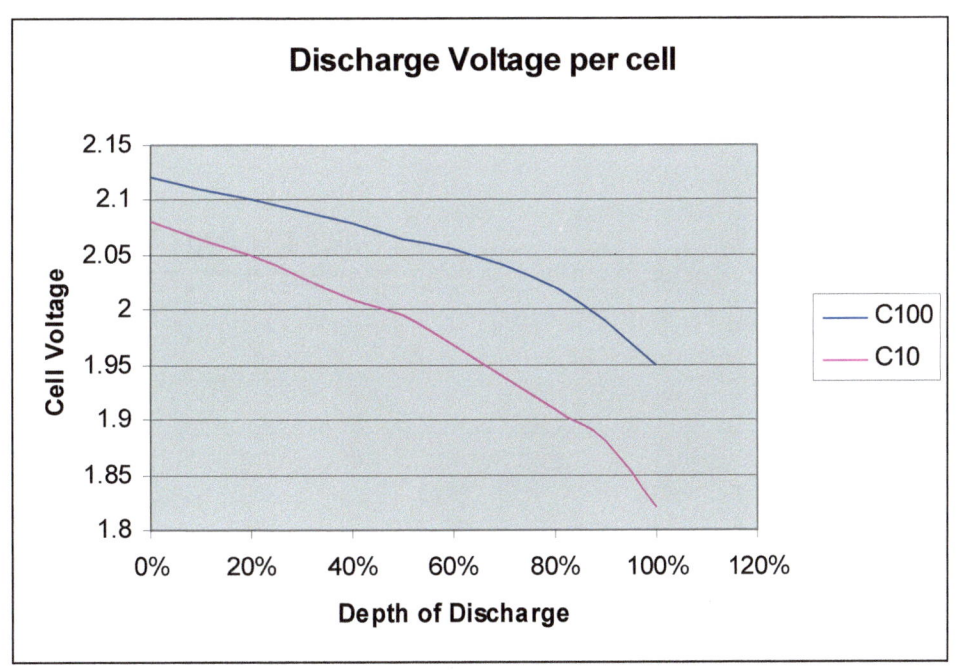

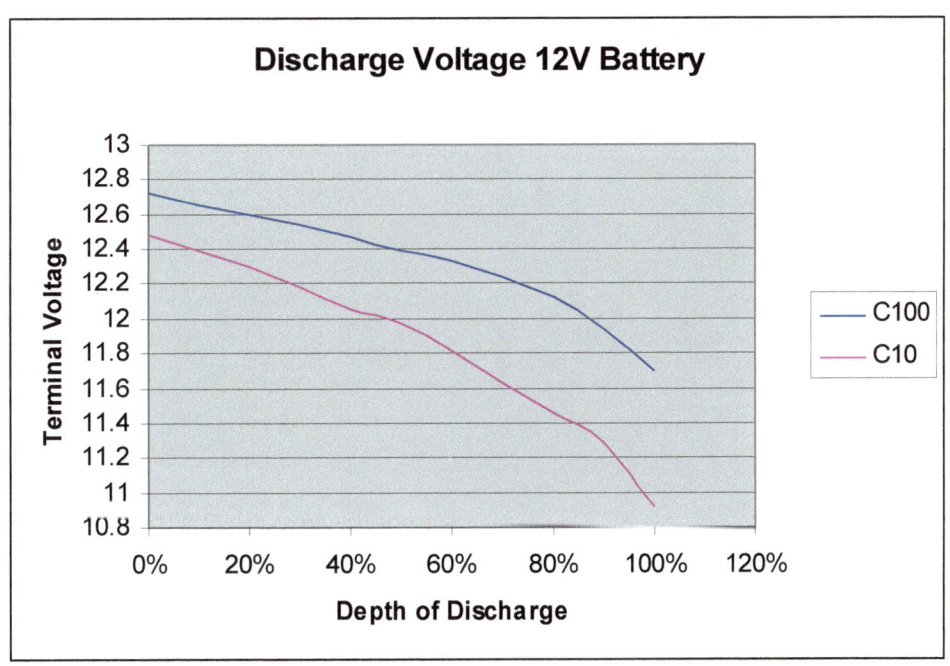

8.3 Appendix 3 – Cable Data

8.3.1 Equivalence and Ratings

CSA mm² (VDE)	AWG	Rated Current (A)
1.0	17	13
1.5	15	17
2.5	13	24
4	11	32
6	9	41
10	7	57
16	5	76
25	3	101
35	2	125
50	1/0	151
70	2/0	192
95	4/0	232
120	5/0	269
150	6/0	300

www.solar-power-answers.co.uk

8.3.2 Voltage drop tables

Cable Voltage drop (V) — Current: 1 Ampere

Cable cross-sectional area (mm^2)

Length (m)	1	1.5	2.5	6	10	16	25	35	50	75	95	120	150
1	0.04	0.03	0.02	0.01	0.00	0.00	0.00	0.00	0.00	0.00	0.00	0.00	0.00
2	0.08	0.05	0.03	0.01	0.01	0.01	0.00	0.00	0.00	0.00	0.00	0.00	0.00
5	0.20	0.13	0.08	0.03	0.02	0.01	0.01	0.01	0.00	0.00	0.00	0.00	0.00
10	0.40	0.27	0.16	0.07	0.04	0.03	0.02	0.01	0.01	0.01	0.00	0.00	0.00
20	0.80	0.53	0.32	0.13	0.08	0.05	0.03	0.02	0.02	0.01	0.01	0.01	0.01
50	2.00	1.33	0.80	0.33	0.20	0.13	0.08	0.06	0.04	0.03	0.02	0.02	0.01
100	4.00	2.67	1.60	0.67	0.40	0.25	0.16	0.11	0.08	0.05	0.04	0.03	0.03

Cable Voltage drop (V) — Current: 10 Ampere

Cable cross-sectional area (mm^2)

Length (m)	1	1.5	2.5	6	10	16	25	35	50	75	95	120	150
1	0.40	0.27	0.16	0.07	0.04	0.03	0.02	0.01	0.01	0.01	0.00	0.00	0.00
2	0.80	0.53	0.32	0.13	0.08	0.05	0.03	0.02	0.02	0.01	0.01	0.01	0.01
5	2.00	1.33	0.80	0.33	0.20	0.13	0.08	0.06	0.04	0.03	0.02	0.02	0.01
10	4.00	2.67	1.60	0.67	0.40	0.25	0.16	0.11	0.08	0.05	0.04	0.03	0.03
20	8.00	5.33	3.20	1.33	0.80	0.50	0.32	0.23	0.16	0.11	0.08	0.07	0.05
50	20.00	13.33	8.00	3.33	2.00	1.25	0.80	0.57	0.40	0.27	0.21	0.17	0.13
100	40.00	26.67	16.00	6.67	4.00	2.50	1.60	1.14	0.80	0.53	0.42	0.33	0.27

Cable Voltage drop (V) — Current: 100 Ampere

Cable cross-sectional area (mm^2)

Length (m)	1	1.5	2.5	6	10	16	25	35	50	75	95	120	150
1	4.00	2.67	1.60	0.67	0.40	0.25	0.16	0.11	0.08	0.05	0.04	0.03	0.03
2	8.00	5.33	3.20	1.33	0.80	0.50	0.32	0.23	0.16	0.11	0.08	0.07	0.05
5	20.00	13.33	8.00	3.33	2.00	1.25	0.80	0.57	0.40	0.27	0.21	0.17	0.13
10	40.00	26.67	16.00	6.67	4.00	2.50	1.60	1.14	0.80	0.53	0.42	0.33	0.27
20	80.00	53.33	32.00	13.33	8.00	5.00	3.20	2.29	1.60	1.07	0.84	0.67	0.53
50	200.00	133.33	80.00	33.33	20.00	12.50	8.00	5.71	4.00	2.67	2.11	1.67	1.33
100	400.00	266.67	160.00	66.67	40.00	25.00	16.00	11.43	8.00	5.33	4.21	3.33	2.67

8.4 Appendix 4 – Example wiring diagrams

8.4.1 12 Volt lighting system

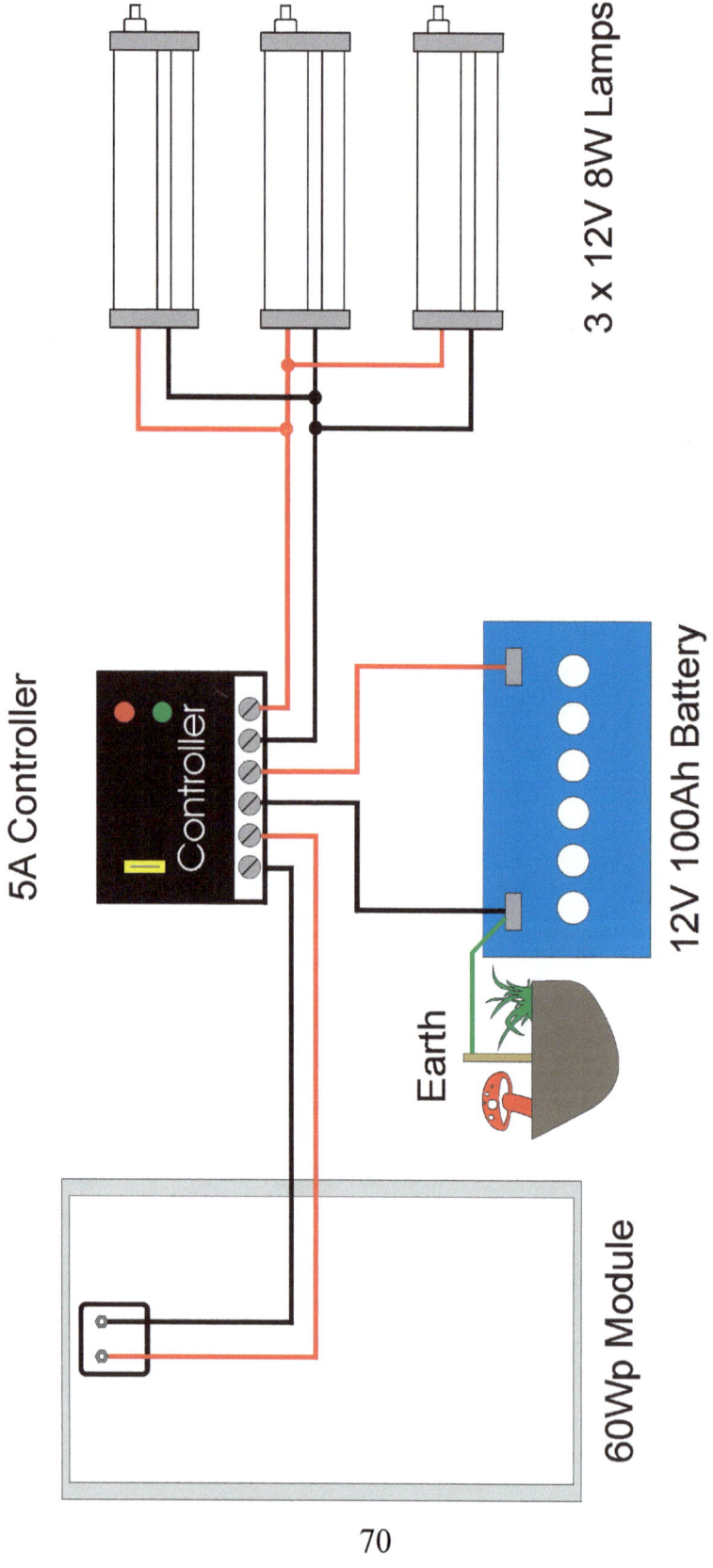

www.solar-power-answers.co.uk

8.4.2 24 Volt inverter system

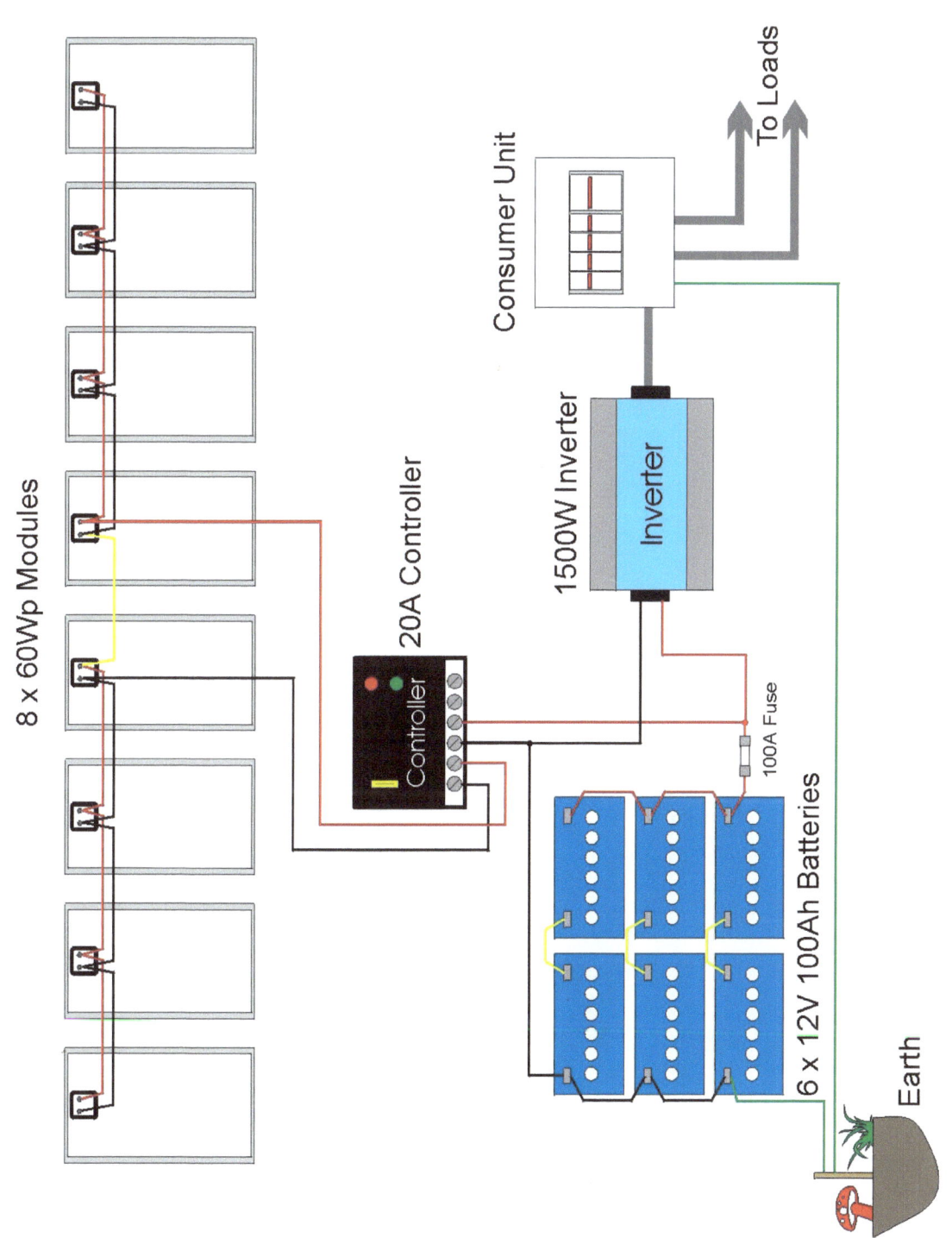

8.4.3 Module wiring for 24 Volt systems

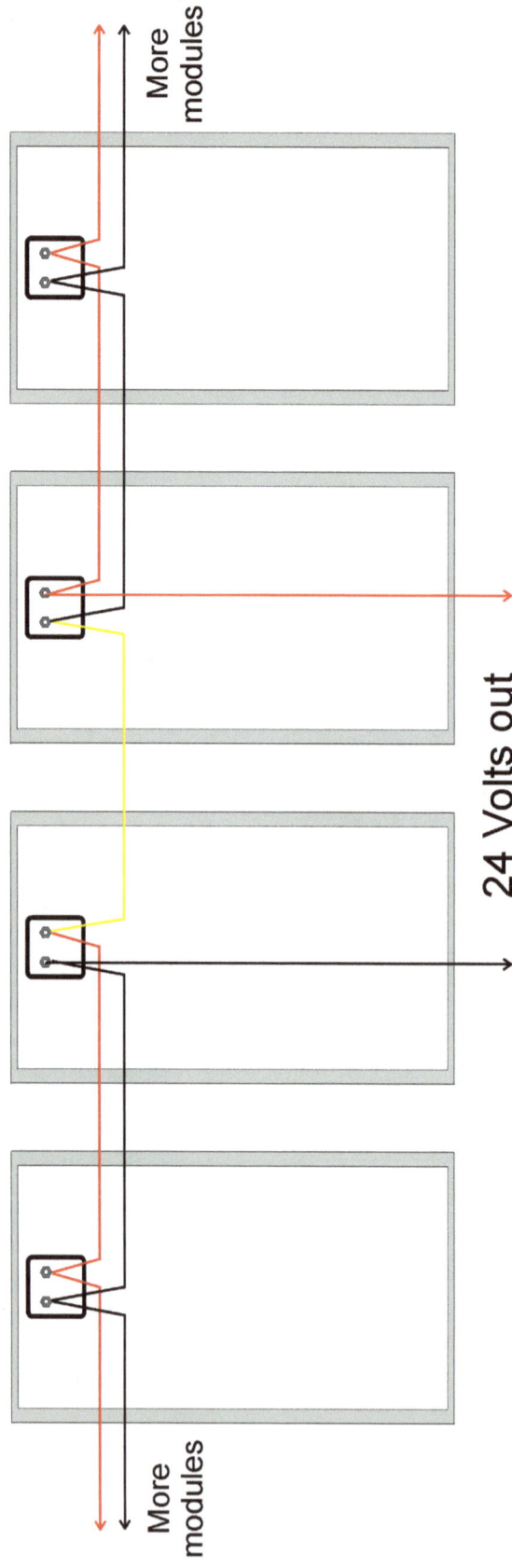

8.4.4 Example Grid-connected system

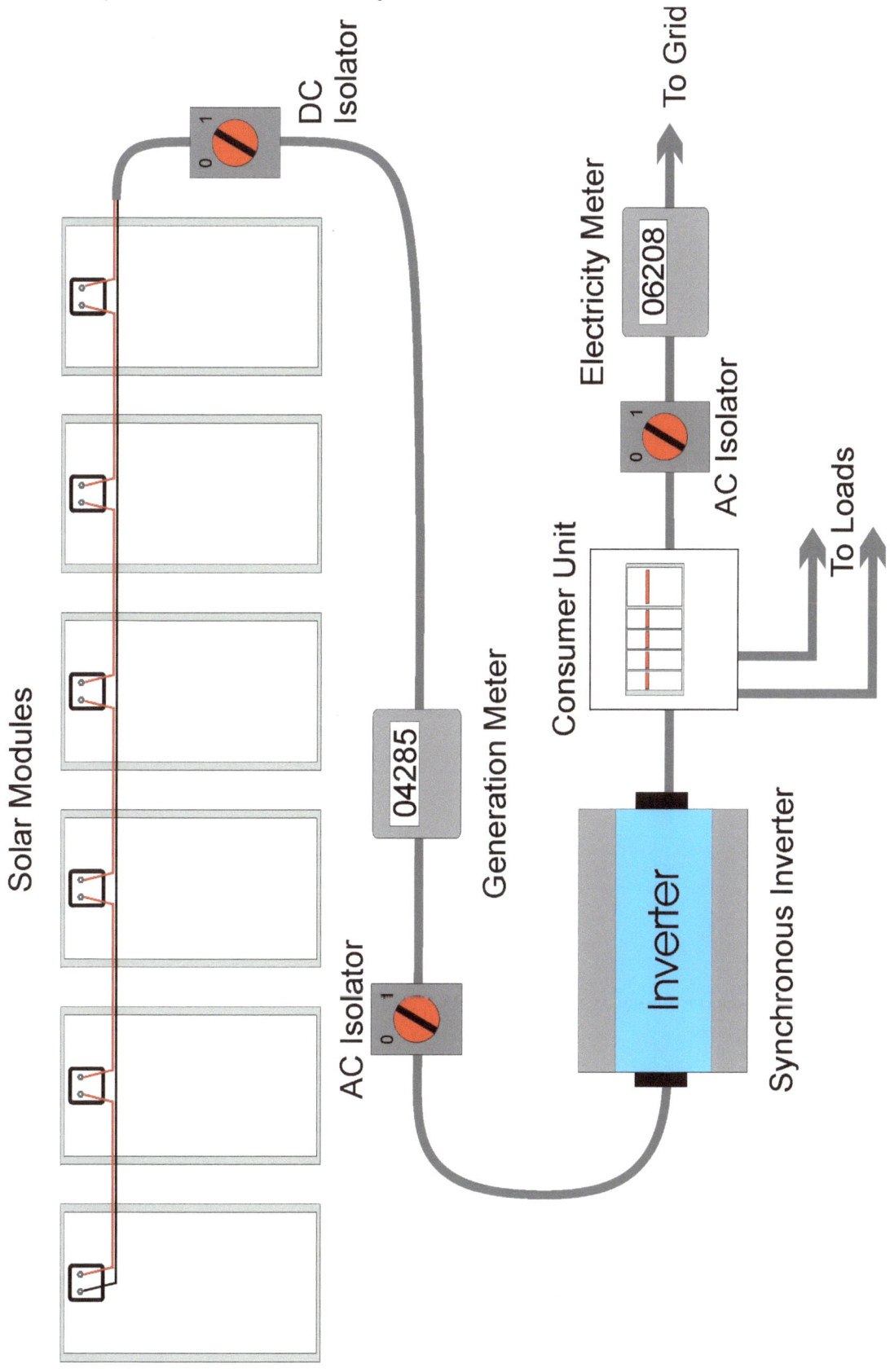

8.5 Appendix 5 – Power ratings of common appliances

Typical power consumption for a range of appliances. Power consumption for 12 Volt versions is given where appropriate.

Appliance	Power Rating (W) 230V Supply	Power Rating (W) 12V Supply
Light (energy saving)	15	10
Television	80	35
Refrigerator	90	50
Washing Machine	3000	-
Dishwasher	3000	-
Fan Heater	2000	-
Central Heating Pump	60	-

www.solar-power-answers.co.uk

www.ingramcontent.com/pod-product-compliance
Lightning Source LLC
Chambersburg PA
CBHW050742180526
45159CB00003B/1314